Thin Films - Deposition Methods and Applications

Edited by Dongfang Yang

Published in London, United Kingdom

Thin Films - Deposition Methods and Applications
http://dx.doi.org/10.5772/intechopen.100701
Edited by Dongfang Yang

Contributors
Katherine Lochhead, Dongfang Yang, Eric Johlin, Elizabeth C. Pastrana, Pierre G. Ramos, Luis A. Sanchez, Juan M. Rodriguez, Marcela Socol, Nicoleta Preda, Carmen Breazu, Oana Rasoga, Md Abdul Majed Patwary, Cícero L.A. Cunha, Tales C. Pimenta, Mariana Amorim Fraga, Macdenis Onyekachi Egbuhuzor, Solomon Chibuzo Nwafor, Chima Umunnakwe, Sochima Egoigwe, Mandla Msimanga, Ning Song, Shuo Deng, Humaira Ghazal, Nadeem Sohail, Karampuri Yadagiri, Tao Wu

ASSISTANT TO THE EDITOR: Katherine Lochhead

First published in London, United Kingdom, 2023 by IntechOpen
IntechOpen is the global imprint of INTECHOPEN LIMITED, registered in England and Wales, registration number: 11086078, 5 Princes Gate Court, London, SW7 2QJ, United Kingdom

British Library Cataloguing-in-Publication Data
A catalogue record for this book is available from the British Library

Additional hard and PDF copies can be obtained from orders@intechopen.com

Thin Films - Deposition Methods and Applications
Edited by Dongfang Yang
p. cm.
Print ISBN 978-1-80356-455-5
Online ISBN 978-1-80356-456-2
eBook (PDF) ISBN 978-1-80356-457-9

For EU product safety concerns:
IN TECH d.o.o., Prolaz Marije Krucifikse Kozulić 3, 51000 Rijeka, Croatia,
info@intechopen.com or visit our website at intechopen.com.

Meet the editor

Dongfang Yang received his Ph.D. in physical chemistry from the University of Guelph (Ontario) in 1995. He joined the National Research Council of Canada in London, Ontario, in 2001 and is now a senior research officer. His current research interests include: laser materials processing; pulsed laser, sputtering and e-beam deposition of thin films; new materials development for energy storage devices; chemical and optical sensor development; and electrochemical studies of organic adsorption and self-assembly monolayers. He is an editor/member of the editorial board for over eleven scientific journals and was listed among the top 2% of most-cited scientists in 2020, according to a recent Stanford study. He received the IAAM Scientist Award and Medal in 2021 and was named a Vebleo Fellow in 2022.

Contents

Preface

Thin films have been employed in many applications as surfaces that possess specific optical, electronic, chemical, mechanical, and thermal properties. This book covers the deposition methods and applications of thin films. It also includes a study of ion beam techniques for depth profiling of multilayer thin films. These ten chapters include original research studies and literature reviews written by experts from the international scientific community.

Chapter 1 presents an overview of various thin film deposition technologies used in the fabrication of photovoltaics. Chapter 2 gives a comprehensive overview of sputtering processes, including the design and basic operations of the sputtering systems, the effects of mass, concentration, energy and angle of incident ions on the ion–matter interaction during the sputtering process, and the benefits, limitations and future trends for sputtering techniques. Chapter 3 presents a very detailed overview of the successive ionic layer adsorption and reaction (SILAR) technique for the deposition of thin films of metal oxides, sulfides, selenides, and tellurides. Chapter 4 explains the fundamentals and applications of ion–matter interactions at MeV energies. It also gives examples of the ion beam analysis (IBA) techniques (Rutherford Backscattering Spectrometry (RBS) and Elastic Recoil Detection Analysis (ERDA)) used to analyze multilayer thin films. Chapter 5 outlines perovskite degradation mechanisms and includes a summary of the progress made to date in the encapsulation of perovskite solar cells, with a particular focus on the most recent and promising advances that employ thin films. Chapter 6 introduces work on the deposition and magnetic property characterization of the bilayer thin film, Ta/CoFeB, deposited by sputtering. It reveals that post-annealing can induce a change in the crystal structure of the bilayer film from amorphous to crystalline as well as alter its damping factor and exchange stiffness. Chapter 7 presents an overview of the fabrication and application of photocatalysts based on thin films of CuO or Fe_2O_3 composited with other semiconducting materials. It also introduces the design and construction of semiconductor–semiconductor heterostructures and gives various examples of α-Fe_2O_3- and CuO-based heterostructures for the photocatalytic degradation of pollutants in wastewater. Chapter 8 describes research on the deposition and characterization of transparent conductive oxide (i.e., ITO and AZO) thin films on glass substrates with or without surface patterns created by ultraviolet nanoimprint lithography. Chapter 9 is a detailed review of the crystalline structures, physicochemical properties, deposition and characterization of AlN thin film as well as the relationship between structure-property and deposition process parameters. Chapter 10 summarizes the fundamentals, advantages and applications of various thin film batteries.

Through these chapters, the reader will gain a better understanding of various thin film deposition techniques and their applications. I gratefully acknowledge the enthusiastic and collaborative contributions of all the authors. Thanks also to Ms. Karla Skuliber, Publishing Process Manager, for her guidance and support in the preparation of this book.

Dongfang Yang, Ph.D.
Senior Research Officer,
Automotive and Surface Transportation,
National Research Council Canada,
Ontario, Canada

Section 1

Thin Film Deposition Methods and Analytical Techniques

Chapter 1

Thin Film Deposition Technologies and Application in Photovoltaics

Ning Song and Shuo Deng

Abstract

Renewable energy will play a critical role in reducing emissions to mitigate climate change. Photovoltaic (PV) is one of the most promising and prominent techniques for electricity generation based on renewable solar energy. Thin films play a critical role in PV in Si and thin film solar cells and solar modules. They can be used as an absorber layer, buffer layer, hole/electron transportation layer, passivation layer, transparent conductive oxide and antireflection coating on solar cells or solar modules. This chapter provides an overview of thin film deposition techniques and applications in photovoltaics and highlights techniques that are currently in use or are promising for mass production.

Keywords: photovoltaic, thin films, Si solar cell, thin films solar cells, solar modules

1. Introduction

In the past two decades, human-induced climate change has increased the frequency and intensity of weather and climate extremes around the globe and has cost countries hundreds of billions of dollars and thousands of lives [1]. The world is facing an energy crisis, and the prices of fossil fuels keep increasing, affecting every household. From personal, national and global perspectives, carbon neutrality is necessary. Renewable energy is a predominant term in carbon-neutral roadmaps for every country, and solar photovoltaics (PV) is currently the most affordable, accessible and prevalent technology. It involves the generation of electricity from sunlight shining through the front cover onto solar cells packaged into a solar module.

As of May 2022, global PV installations have reached 1 TW. In the PV market, crystalline-Si (c-Si) solar cells account for 95% and thin film solar cells account for 5% [2]. Thin films (<1um) have an important role in Si solar cells, thin film solar cells and solar modules as absorber, passivation, buffer, electron/hole transport and antireflection coating (ARC) layers on solar cells and modules. Thin films can range from single crystal to amorphous, fully dense to less than fully dense, pure to impure and thin to thick [3].

Any new technology adapted to PV would be suitable for mass production at a reasonable cost. Meanwhile, the industry is shaping itself in the direction of higher conversion efficiency and lower cost, hence new materials and technologies are always under investigation for companies to keep their products competitive.

Therefore, it is significant to evaluate the scalability and cost-effectiveness of fabrication technologies.

The following benefits of thin films make them naturally attractive for the industry [3]:

- Conservation of scarce materials,
- Production of nanostructured coatings and nanocomposites,
- Ecological considerations—a reduction of effluent output and power consumption,
- Improved functionality of existing products,
- Solution to previously unsolved engineering problems by adding functional thin layers and
- Creation of entirely new and revolutionary products such as thin film solar cells.

In materials science, thin film technologies can be divided into gaseous state, solution state and molten or semi-molten state, where gaseous and solution states produce films with a thickness level of nanometers (nm) and Angstroms (Å), which are at the desired level of controllability for photovoltaics. Considering the accessibility and cost, the main thin film deposition techniques used in photovoltaics are physical vapor deposition (PVD), chemical vapor deposition (CVD), chemical solution deposition and sol–gel [3].

2. Crystalline silicon solar cells

As mentioned above, c-Si is dominating the PV industry with a market share of 95%. In c-Si, thin film technologies are commonly applied to fabricate anti-reflection coatings (to reduce surface reflection loss) and passivation layers (to reduce carrier or surface recombination) during c-Si solar cell manufacturing. **Figure 1** shows schematics and fabrication flow steps for mainstream c-Si solar cells: a. Al back-surface field (Al-BSF) solar cell; b. passivated emitter and rear solar cell (PERC); c. n-type solar cell with a tunnel oxide passivating contact (TOPCon); d. silicon heterojunction solar cell (SHJ) contacted on both sides with intrinsic and doped bilayers at front and rear, respectively, and indium tin oxide (ITO). It can be seen that the deposition of ARC/passivation layers is a key step among all four crystalline silicon solar cell configurations.

The technique mainly used for these ARC/passivation layers is CVD, in particular, PECVD, LPCVD and ALD. PVD-sputtering is the main technique that is used for ITO layers in SHJ solar cells. The distinguishing feature between PVD and CVD is the states of vapor. In PVD, the vapor is made up of atoms and molecules that simply condense on the substrate; in CVD, the vapor undergoes a chemical reaction on the substrate which results in a thin film [3]. Films produced by CVD generally have better quality in terms of very high purity and density and better coverage on rough surfaces than those produced by PVD methods, although the process usually involves toxic and/or corrosive gases [7].

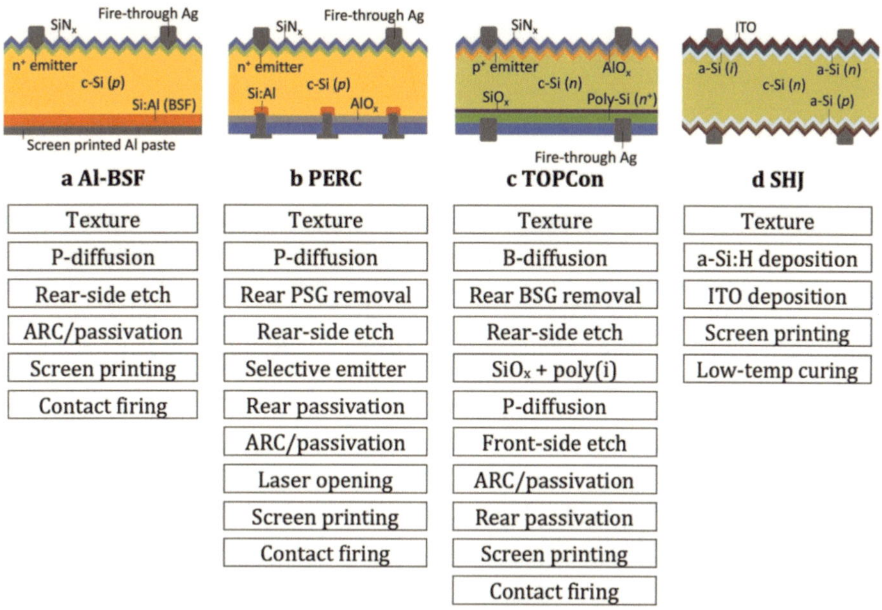

Figure 1.
Schematics and fabrication flow for a. Al back-surface field (Al-BSF) solar cell; b. localized rear contacts in the passivated emitter and rear cell (PERC); c. n-type solar cell with a tunnel oxide passivating contact (TOPCon); d. silicon heterojunction solar cell (SHJ) contacted on both sides with intrinsic and doped bilayers (i/n and i/p at front and rear, respectively) and indium tin oxide (ITO) [4–6].

2.1 PECVD

Plasma-enhanced chemical vapor deposition (PECVD) is one of the most commonly used methods to deposit thin films in c-Si solar cell manufacturing. The current fabrication process involves PECVD deposited silicon nitride (SiN_x) used as a front side anti-reflection coating being applied to Al-BSF, PERC and TOPCon solar cells. PECVD SiN_x:H stacked with PECVD or atomic layer deposition (ALD) deposited AlO_x provides excellent passivation and is involved as a standard step in the PERC and TOPCon fabrication processes [8, 9]. Amorphous silicon (a-Si) is another thin film material with excellent passivation property that is commonly deposited by PECVD [10]. **Figure 2** shows the schematics of a PECVD reactor. The PECVD deposition

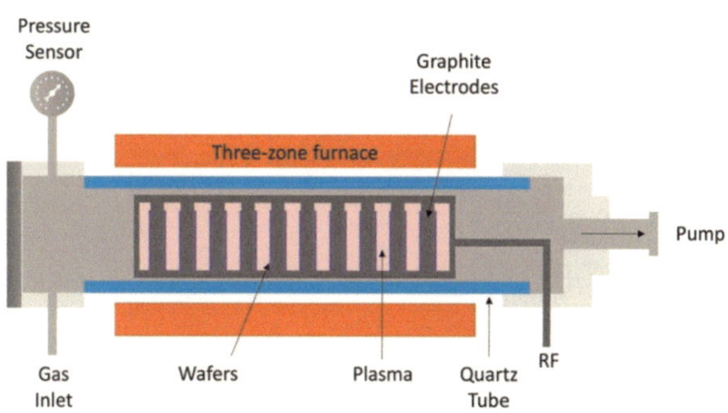

Figure 2.
Schematic of a PECVD reactor.

process normally takes place at a temperature range of 150–400°C and uses RF or DC-generated plasma to initiate the reactant gases into reaction.

2.2 LPCVD

Compared with PECVD, low-pressure chemical vapor deposition (LPCVD) is another CVD thin film deposition technique but with a higher deposition temperature, lower deposition pressure and typically lower deposition rate. It uses heat to initiate a reaction of a precursor gas on the solid substrate. This reaction at the surface forms a solid phase material. The reactor is kept at low pressure to suppress any unwanted gas phase reactions, which also increases the uniformity. The temperatures can range from 400 to 900°C depending on the process and the reactive gases being used. **Figure 3** shows a schematic of an LPCVD reactor. LPCVD deposited films are typically more uniform, have fewer defects and exhibit better step coverage than films produced by PECVD.

LPCVD dominates in producing poly-Si for TOPcon solar cells in the PV industry. However, the technical difficulties of LPCVD are also required to be overcome to reduce the cost, in terms of higher deposition temperature, lower deposition rate and wrap-around deposition issues which increase the energy consumption for production [5, 11].

Basic CVD (PECVD and LPCVD) chemical reactions with silane gas (SiH_4) are shown below:

Silicon nitride (SiN_x):

$$SiH_4 + NH_x \rightarrow SiN_x\ (+H_2).$$

$$\text{Or}\ SiH_4 + N \rightarrow SiN_x\ (+H_2).$$

Silicon oxide (SiO_x):

$$SiH_4 + N_2O \rightarrow SiO_x\ (+H_2 + N_2).$$

Silicon oxynitride ($SiON_x$):

$$SiH_4 + N_2O + NH_3 \rightarrow SiON_x\ (+H_2 + N_2).$$

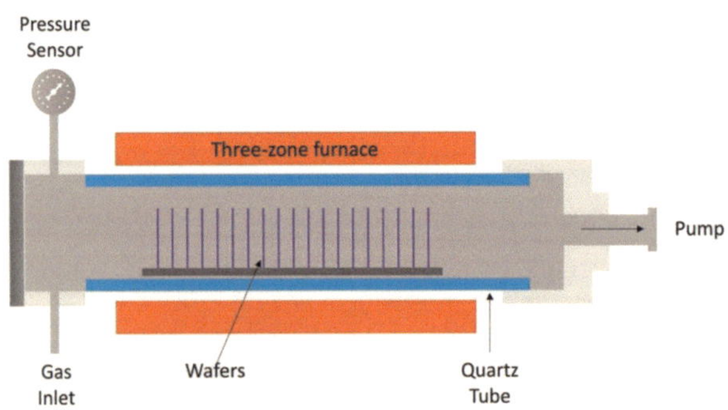

Figure 3.
Schematic of a PECVD reactor.

Hydrogenated amorphous silicon (a-Si:H):

$$SiH_4 \rightarrow Si\ (+H_2).$$

Silicon carbide (SiC_x):

$$SiH_4 + CH_x \rightarrow SiC_x\ (+H_2).$$

2.3 ALD

ALD is a widely used deposition technique in the field of solar cells, energy storage, catalysis and semiconductor technology. ALD deposition for AlO_x has been successfully employed by the PV industry for its excellent film quality and process economics [12–14].

In ALD, thin films are built up in cycles, in which the surface is exposed to various vapor or gas-phase species in alternating, separated doses. One cycle typically involves using the precursor dose as the first half-cycle and the co-reactant as the second half-cycle. In each cycle, a sub-monolayer of material is deposited. The precursor molecules and co-reactants react neither with themselves nor with the surface groups created. The products generated during the gaseous reactions, as well as any unreacted precursor or co-reactant molecules, are removed from the ALD reactor in the purge and/or pump steps. This is necessary to avoid reactions between the precursor and co-reactant molecules directly in the gas phase or on the surface, as this could lead to an undesired CVD component. The various steps in a typical ideal ALD cycle are shown in **Figure 4**. As shown, a typical cycle consists of four steps: (i) a precursor dosing step; (ii) a purge and/or pump step; (iii) a co-reactant step, where a small molecule is typically involved, such as water vapor; and (iv) a purge and/or pump step. This figure shows a schematic illustration of the self-limiting surface reactions during the two half-cycles, as well as the saturation of the surface coverage in every step of the cycle. The saturation of both half-cycles leads to a characteristic amount of growth per cycle (GRC) [15].

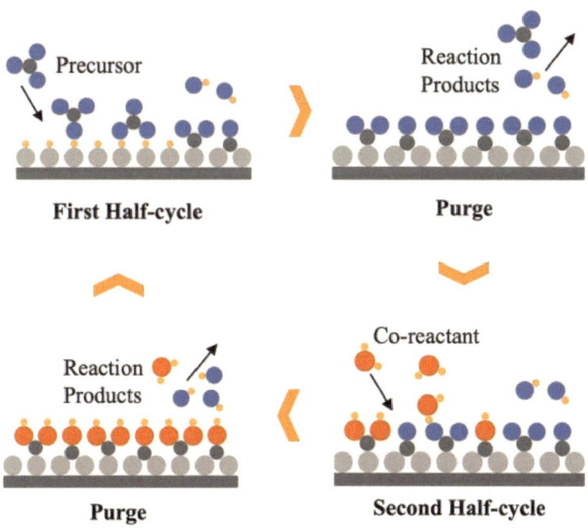

Figure 4.
Illustrations of the ALD cycle process.

3. Thin film solar cells

Even though the photovoltaic module market is dominated by crystalline silicon wafer-based technologies, thin film solar cells have the advantages of being lightweight, flexible, transparent and high temperature and radiation tolerant, making them competitive alternatives for c-Si in applications such as wearable devices, space applications, building-integrated PV and vehicle integrated PV. Thin film solar cells also avoid the massive energy consumption in the fabrication of high-purity silicon ingots in c-Si manufacturing. Moreover, thin film solar cells can be fabricated on Si solar cells to form tandem solar cells, enabling higher conversion efficiency.

Figure 5 shows three different configurations for CdTe, perovskite and CuInGaSe (CIGS) solar cells. The basic structure of CdTe and CIGS (same as Cu_2ZnSnS_4 (CZTS)) solar cells includes a window layer (TCO), buffer layer, absorber layer and back contacts and substrates, while perovskite solar cells have an n-i-p structure with extra electron/holes transport layers. Different layer stacks are coated on the substrate with thin-film coating methods using either solution-based processes like a chemical bath or an ink-like coating procedure, or using vacuum-based processes like thermal evaporation or sputtering [16].

3.1 Solution processing

3.1.1 Spin coating

Since it was first reported in 2009, the power conversion efficiency of perovskite solar cells has increased from 3.8–25% (lab scale) within 13 years. The emerging perovskite solar cells are under massive investigation in both research and industry due to their low-cost and high efficiency, as well as their use in Si-based tandem solar cells with further enhanced efficiency. Solution processing is the traditional fabrication method for perovskite solar cells. Spin coating is the most common fabrication approach used in the laboratory, and it has the advantages of a simple process, with no requirement for expensive and complex vacuum systems. This method could also introduce additives into the perovskite precursor solution to improve device performance [17]. **Figure 6** shows the schematics of the spin coating method for perovskite

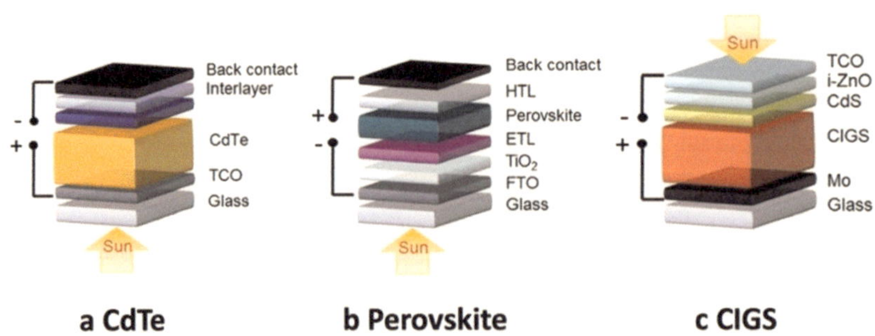

Figure 5.
(a) Superstrate configuration for CdTe cells; (b) perovskite cells in the n-i-p configuration; c. substrate CIGS cell configuration. Reproduced from [16] with permission.

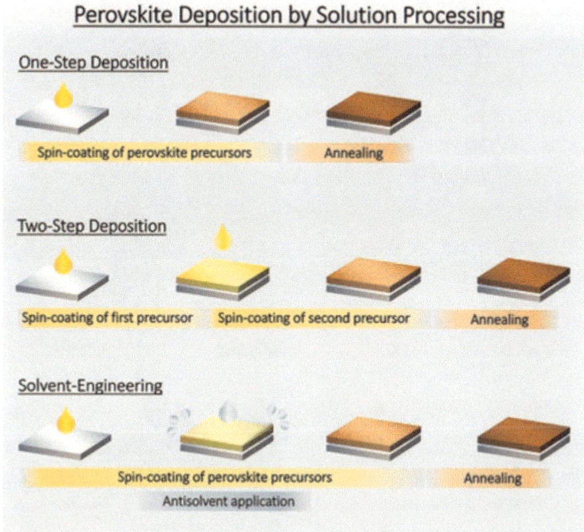

Figure 6.
Schematic illustration of the deposition of a perovskite layer by spin-coating and post annealing. In the one-step method, all perovskite precursors are mixed in a single solution, which is deposited onto the substrate; In the two-step approach, the precursors is deposited first, followed by the deposition of a second precursor and thermal annealing; solvent engineering approach is to deposit all perovskite precursors in a single step, during which an antisolvent is applied triggering the crystallization of the perovskite film. Reproduced from [17] with permission.

fabrication. The fabrication of perovskite absorber can be achieved by one-step deposition, two-step deposition or solvent engineering methods with spin-coating deposition and post annealing processes, as shown in **Figure 6**. However, the spin-coating method is generally not suitable for mass production.To translate such PV technology from the laboratory to industrial-scale manufacturing, other techniques need to be explored.

3.1.2 Slot-die coating

One of the most promising techniques for achieving large-scale (roll to roll) perovskite solar cell production is slot-die coating [18–20].

Slot-die coating can achieve highly precise control of material usage and results in very low waste levels of ink compared to other deposition methods such as spin coating or spray and screen printing. **Figure 7** shows the schematics of slot-die coating. By adjusting the ink flow pumped to the die coating head by syringe and the substrate speed, fine control of the deposited film thickness can be achieved, from tens of nanometers to tens of microns. An air knife with nitrogen gas flow would help with drying the perovskite ink, and heating could also be applied for better crystallization [18, 21].

The film formation process can also be controlled by the choice of precursor, solvent and additive. **Figure 8a** shows a schematic of slot-die coating setup with two inks and the influence of the strongly coordinating solvent dimethyl-sulfoxide (DMSO) as an additive in 2-methoxy-ethanol (2-ME) based perovskite ink on film formation. Adding a limited amount of DMSO (11.77 mol%) leads to a denser

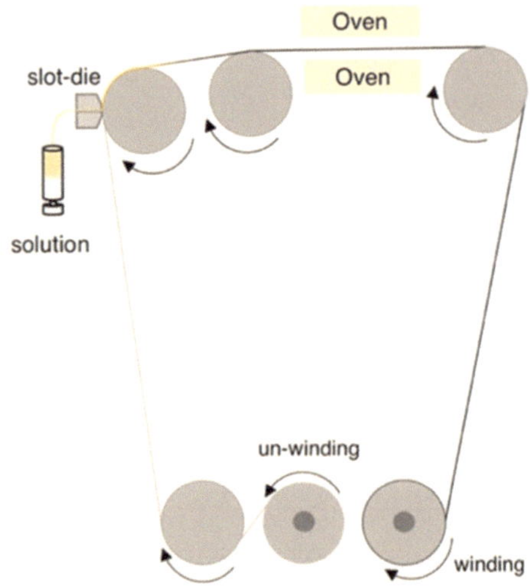

Figure 7.
Schematic of roll-to-roll slot-die setup. Reproduced from [20] with permission.

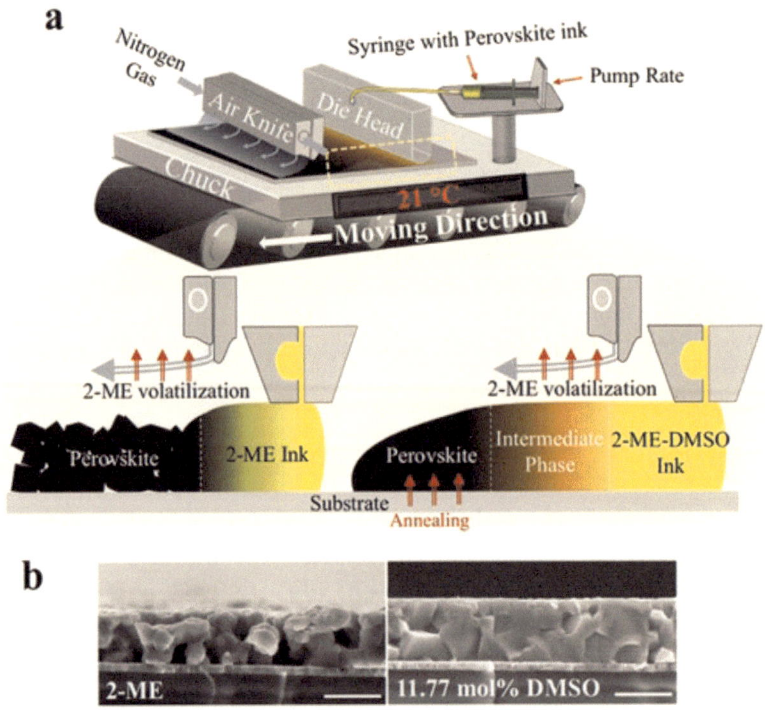

Figure 8.
(a) Schematic of slot-die coater setup with 2-ME ink and 2-ME-DMOS ink. (b) SEM cross-sectional images of the coated films without and with DMOS. Reproduced from [22] with permission.

thin-film without pinholes and large columnar crystallites as shown in the SEM cross-section images in **Figure 8b** [22]. Research on the film morphology and optimization of perovskite ink and chemicals is still ongoing.

3.1.3 Inkjet printing

Inkjet printing is another promising large-scale production method for perovskite solar cells, as it offers several favorable properties. Ink-jet printing of perovskite can achieve high-resolution film formation with customized patterns under ambient conditions.

A schematic of inject printing is shown in **Figure 9**. Inkjet printing of perovskite is carried out by a drop-on-demand (DOD) printing approach, which by definition, enables the generation of a single droplet when required, hence enabling high precision of material usage. The location of film formation can be finely controlled by motions of the inkjet printhead nozzles and the substrate. The ejection of printing materials (perovskite ink) is forced out of the nozzle by regular pressure pulses caused by contractions of ink chamber volume in the nozzles. The pressure pulse can be generated by either the mechanical deformation of a piezoelectric transducer or the collapse of thermal bubbles which involves resistive localized heating in the ink chamber. Due to the technical challenge of generating vapor bubbles in high-vapor pressure fluid, and the fact that a piezoelectric inkjet printing system can control droplet size and velocity by simply adjusting the actuation pulse, piezoelectric DOD inkjet printing would be more suitable for large-scale production of perovskite solar cells. **Figure 9** demonstrates the schematic of an inkjet printing nozzle, which an industry-scale printhead would have several hundreds of, enabling rapid and high-resolution printing at a low cost [23, 24].

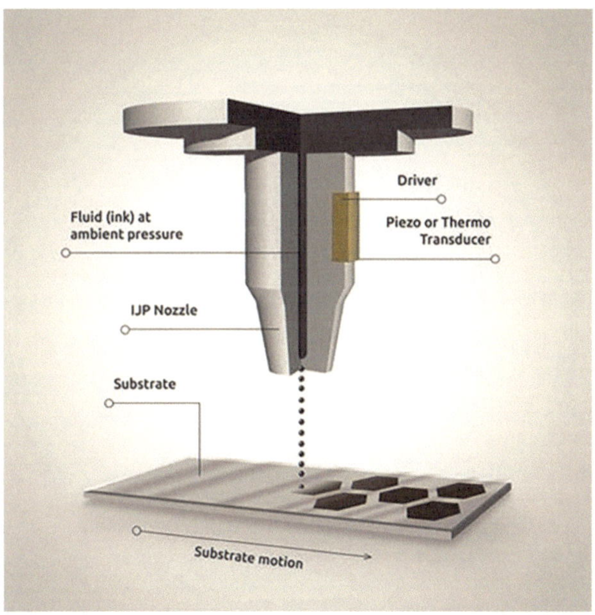

Figure 9.
The schematic of an inkjet printing nozzle (Saule technologies). Reproduced from [23] with permission.

3.2 Vacuum-based processing

3.2.1 Evaporation

Among vacuum-based processing techniques, evaporation is a widely used deposition process for the formation and growth of thin films in PV. The process is beneficial in a contemporary environment and extensively applicable in the laboratory and industrial manufacturing for the deposition of thin films. A schematic of the evaporation coating is shown in **Figure 10a**.

The basic sequential steps for the evaporation process are given below:

- A vapor is created from subjecting the target material to a very high temperature by subliming or boiling (thermal) or bombarding with a beam of electrons (E-beam).
- The ejected vapor from the target material is transported to the substrate through a vacuum.
- Condensation of the vapor takes place to form a solid thin film on the surface of the substrate and further repeats of the deposition cycles result in thin film growth and nucleation.

The evaporation process has been reported to be performed using different configurations, including molecular beam epitaxy, reactive evaporation and activated reactive evaporation [3].

Snaith and his team first demonstrated that perovskite can also be deposited by thermal evaporation, reaching an efficiency of 15.4% in 2013 [25]. A recent study by

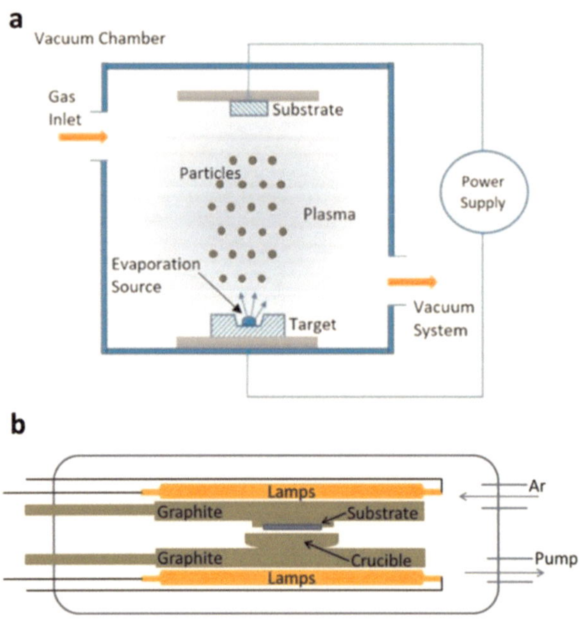

Figure 10.
Schematics of (a) evaporation coating methods. Reproduced from [36] with permission. (b) Close space sublimation (CSS) system. Reproduced from [37] with permission.

Bruno and co-workers demonstrated thermally evaporated perovskite mini-modules with an active area of 21 cm^2 and an impressive efficiency of 18.13% [26]. These results are superior to all other reports of large-area devices in the literature, among which the highest is a blade-coated perovskite solar cell with a maximum efficiency of 16.4% [27]. Notable progress was made in 2016 by Momblona *et al.*, who demonstrated a fully evaporated planar perovskite solar cell with an efficiency above 20%. It is worth noting that the charge carrier-selective layers, the MAPbI3 absorber and the metal electrode were all sequentially evaporated, demonstrating an all vacuum-based process for the first time [28].

Thermal evaporation offers the unique possibility of depositing multilayers of perovskite materials. Moreover, the fact that perovskite layers can be formed without the need for thermal annealing makes thermal evaporation particularly suitable for flexible optoelectronics application, in which low processing temperatures are required. Finally, this process avoids the use of the toxic solvents and allows the deposition of perovskite films without the risk of damaging the underlying layers of tandem devices [17].

Thermal evaporation is also employed in fabrication processes for commercial CIGS and CdTe solar cells [29–32]. CIGS absorbers are usually fabricated by thermal co-evaporation of the constituents, using the so-called 3-stage process. During this process, In, Ga and Se are evaporated in the 1st and 3rd stages, while Cu and Se are deposited in between, leading to the so-called double gradient of the In and Ga concentrations [33].

Close-spaced sublimation (CSS) is the fastest and simplest deposition process for both the CdTe and CdS semiconductors used in CdTe thin film solar cells, permitting high-speed in-line production [34]. The current CdTe thin film solar cell record is 21% on an approximately 1cm^2 cell on glass made by First Solar [35].

Figure 10b shows a schematic of a CSS system, in which the CdTe is deposited at a pressure between 1 and 100 mbar in argon or nitrogen. The substrate and crucible are kept a few centimeters apart for vapourization of CdTe granulate and condensation (crystallization) on the substrate. Because the substrate is kept at a temperature range from 450–600°C for high-quality crystallization, a relatively high pressure is applied to suppress re-evaporation of the material [34, 37, 38].

3.2.2 Sputtering

Sputtering is another widely used PVD technique in PV that can be upscaled. In sputtering processes, a magnetron is positioned near the target. The ionic gas is introduced in an accelerated way into the vacuum chamber, blasting the target, releasing atomic-sized particles to be deposited, which will be violently projected onto the substrate. A schematic of sputtering deposition is shown in **Figure 11a**. Sputtering deposition has become a generic name for a variety of sputtering processes. These processes are named based on their source and the orientation of the process. Variants of sputtering include diode sputtering (cathode or radio frequency), reactive sputtering, bias sputtering, magnetron sputtering and ion-beam sputtering with a DC or RF power source [3]. Because of the mechanism nature of sputtering, it is ideal for the deposition of doped materials with multiple target sources. With gas inlets, the sputtering process could be used for the deposition of thin film metal or nonmetal oxides, and with hydrogen gas, hydrogenated thin films as well. Sputtering enables the production of smooth surfaces using lower temperatures, presenting excellent mechanical and tribological properties and having very good adhesion to the main materials used as substrate.

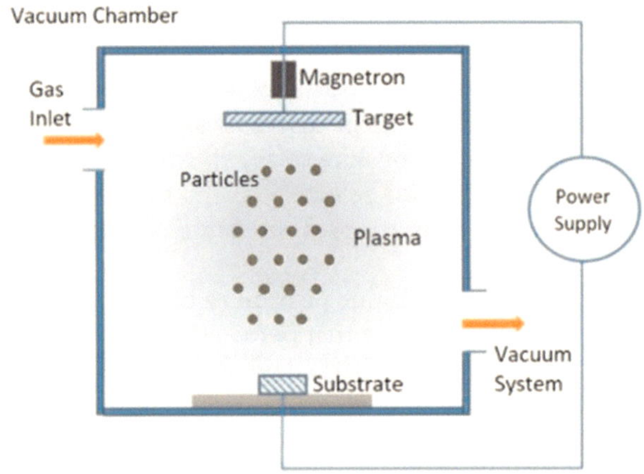

Figure 11.
Schematic of sputtering deposition. Reproduced from [36] with permission.

Sputtering has been used in the deposition of CIGS and CZTS absorber layers [39, 40]. A record efficiency of 23.35% in CIGSSe thin film solar cells was achieved with a sputtered absorber layer [40]. Sputtering is also commonly used for TCO, buffer layers and ETL/HTL with materials such as ITO [41], FZO [42], MoO_x [43], AZO [44], TiO_x [45], and other metal oxides as interlayers, passivating thin film solar cells [46–48]. A commercialized all-sputtering system was developed by Midsummer for CIGS production with process sequence completed in different sputtering chambers for diffusion barrier, absorber, buffer, window and TCO layer deposition [49]. The sputtering method could also be employed for the deposition of poly-Si contacts and passivation for heterojunction solar cells [50–53].

3.2.3 MOCVD (MOVPE)

III–V thin film solar cells are widely used in aerospace applications, due to the high energy conversion rate, wide operating temperature range and high radiation resistance [54]. The record efficiencies for III–V multijunctions are 38% for a five-junction cell (bonded) and 37.9% for InGaP/GaAs/InGaAs solar cells. The record efficiency for a GaAs single-junction cell is 25.1%, held by Alta Devices [35]. One key deposition method used in III–V thin film solar cell fabrication is metalorganic chemical vapor deposition (MOCVD), also called metalorganic vapor-phase epitaxy (MOVPE). Metal–organic CVD (MOCVD) is a CVD process for growing epitaxial films and is done by flowing precursor gases over the substrate. In III–V semiconductors, the metallic element is carried by an organic gas such as $Ga(CH_3)_3$) and $In(CH_3)_3$ along with AsH3 or phosphine (PH_3). The gases are allowed to decompose due to pyrolysis on the heated substrate surfaces to produce the desired film. A schematic is shown in **Figure 12**. Commercially available MOCVD tools are designed to produce traditional III–V semiconductor devices for electronic and optoelectronic applications. Relative to solar cells, these devices have complex layered structures that require extremely precise control of thickness, composition and doping profiles, and each fabrication typically takes multiple hours to complete. The simpler structure and relatively wide process windows of solar cells present an opportunity to use correspondingly simpler

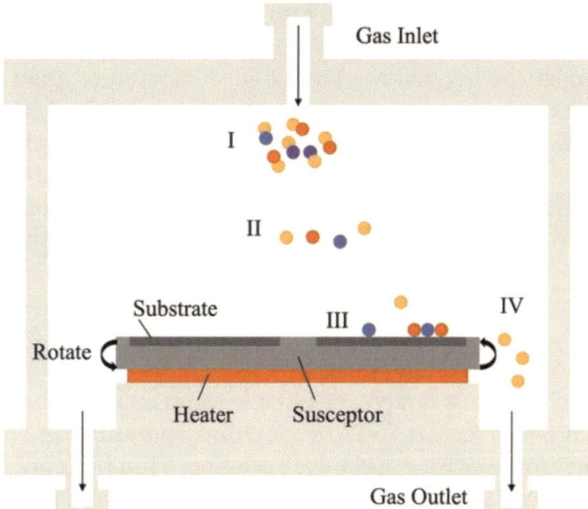

Figure 12.
MOCVD reactor basic scheme and fundamental working principle: I—Precursor, II—Cracking, III—Deposition and IV—Removal of residual gases.

and less expensive equipment. With customized MOCVD, combining automated substrate loading and unloading, fast temperature ramping, high growth rate and the elimination of pressure cycling, the fabrication process time could be reduced to 15 min, which makes a significant contribution to fabrication cost reduction [55].

4. Solar modules

Another important application of thin films in PV is the antireflection coating (ARC) on the surface of solar glass where the light first reaches the solar panels. Currently, single-layer antireflection coated solar glass has a dominant market share of 95% compared to glass with other coatings or no coating, for Si PV modules [2]. This ARC results in an efficiency gain of 2–3%; that is, 2–3% more light can enter the solar modules to be

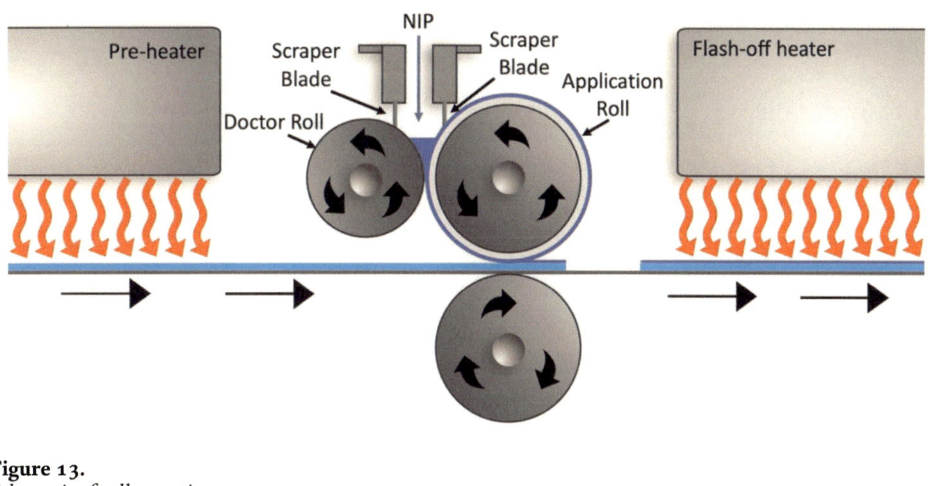

Figure 13.
Schematic of roller coating process.

absorbed by the solar cells and converted into electricity. The most common PV ARC consists of a ~ 100 nm single layer of nano-porous silica deposited onto the solar glass cover via sol–gel roller coating, followed by a high-temperature sintering and tempering process. The roller coating is one type of sol–gel method that is scalable and low cost. A schematic of a roller coating process is shown in **Figure 13**. Other chemical methods such as slot-die coating, dip coating, spin coating, spray coating, as well as PVD methods like sputtering multilayer ARC are also under investigation.

5. Conclusion

This chapter reviewed thin film deposition techniques in the PV industry and research and highlighted those which are suitable for mass production including vacuum-based techniques such as CVD, evaporation, sputtering and non-vacuum based like slot-die coating, roller coating and inkjet printing. Thin films are important components for solar cell and solar module devices which enable high solar energy to electricity conversion efficiency. The increasingly enormous deployment of PV and rapid development of the PV industry has placed higher requirements for fabrication technologies and equipment in terms of good film properties and low cost for mass production.

Acknowledgements

The authors acknowledge the support from the Australian government through the Australian Renewable Energy Agency (ARENA) and the Australian Centre of Advanced Photovoltaics (ACAP, Grant No. R5-NS01) and UNSW Career Advanced fund. The authors also acknowledge Dr. Kaiwen Sun, Dr. Udo Römer, Mr. Xinyuan Wu, Dr. Chang Yan, Dr. Lei Shi and Dr. Jianghui Zheng and Dr. Jizhong Yao for valuable discussion and advice.

Conflict of interest

The authors declare no conflict of interest.

Author details

Ning Song* and Shuo Deng
University of New South Wales, Sydney, Australia

*Address all correspondence to: n.song@unsw.edu.au

References

[1] Jackson RB et al. Global fossil carbon emissions rebound near pre-COVID-19 levels. Environmental Research Letters. 2022;**17**(3):031001

[2] ITRPV. International Technology Roadmap for Photovoltaic. Twelfth ed. 2021

[3] Oluwatosin Abegunde O et al. Overview of thin film deposition techniques. AIMS Materials Science. 2019;**6**(2):174-199

[4] Ballif C et al. Status and perspectives of crystalline silicon photovoltaics in research and industry. Nature Reviews Materials. 2022;**7**(8):597-616

[5] Chen D et al. 24.58% total area efficiency of screen-printed, large area industrial silicon solar cells with the tunnel oxide passivated contacts (i-TOPCon) design. Solar Energy Materials and Solar Cells. 2020;**206**:110258

[6] Deng W et al. 22.61% efficient fully screen printed PERC solar cell. In: 2017 IEEE 44th Photovoltaic Specialist Conference (PVSC). IEEE; 2017

[7] Green J. Table Comparison: Physical vapor deposition Vs. Chemical Vapor Deposition. 2021. Available from: https://www.sputtertargets.net/blog/table-comparison-physical-and-chemical-vapor-deposition.html

[8] van de Loo BW et al. On the hydrogenation of poly-Si passivating contacts by Al2O3 and SiNx thin films. Solar Energy Materials and Solar Cells. 2020;**215**:110592

[9] Gao T et al. An industrially viable TOPCon structure with both ultra-thin SiOx and n+−poly-Si processed by PECVD for p-type c-Si solar cells. Solar Energy Materials and Solar Cells. 2019;**200**:109926

[10] Neuhaus D-H, Münzer A. Industrial silicon wafer solar cells. Advances in Optoelectronics. 2007;**2007**:24521

[11] Chen Y et al. >25% large-area industrial silicon solar cell: Learning from history and future perspective. In: 36th European Photovoltaic Solar Energy Conference and Exhibition. 2019

[12] Huang H et al. 20.8% industrial PERC solar cell: ALD Al2O3 rear surface passivation, efficiency loss mechanisms analysis and roadmap to 24%. Solar Energy Materials and Solar Cells. 2017;**161**:14-30

[13] Hoex B et al. Ultralow surface recombination of c-Si substrates passivated by plasma-assisted atomic layer deposited Al 2 O 3. Applied Physics Letters. 2006;**89**(4):042112

[14] Hossain MA et al. Atomic layer deposition enabling higher efficiency solar cells: A review. Nano Materials Science. 2020;**2**(3):204-226

[15] Kuech T. Handbook of Crystal Growth: Thin Films and Epitaxy. Elsevier; 2014

[16] Powalla M, et al. Thin-film solar cells exceeding 22% solar cell efficiency: An overview on CdTe-, Cu(In,Ga)Se2-, and perovskite-based materials. Applied Physics Reviews. 2018;**5**(4):041602

[17] Vaynzof Y. The future of perovskite photovoltaics—thermal evaporation or solution processing? Advanced Energy Materials. 2020;**10**(48):2003073

[18] Patidar R et al. Slot-die coating of perovskite solar cells: An overview.

Materials Today Communications. 2020; **22**:100808

[19] Yang Z et al. Slot-die coating large-area formamidinium-cesium perovskite film for efficient and stable parallel solar module. Science Advances. 2021;**7**(18): eabg3749

[20] Dou B et al. Roll-to-roll printing of Perovskite solar cells. ACS Energy Letters. 2018;**3**(10):2558-2565

[21] Li H et al. Recent progress towards roll-to-roll manufacturing of perovskite solar cells using slot-die processing. Flexible and Printed Electronics. 2020; **5**(1):014006

[22] Li J et al. 20.8% slot-die coated MAPbI3 perovskite solar cells by optimal DMSO-content and age of 2-ME based precursor inks. Advanced Energy Materials. 2021;**11**(10):2003460

[23] Forgacs D, Wojciechowski K, Malinkiewicz O. Perovskite photovoltaics: From laboratory to industry. In: High-Efficient Low-Cost Photovoltaics. Springer; 2020. pp. 219-255

[24] Peng X et al. Perovskite and organic solar cells fabricated by inkjet printing: Progress and prospects. Advanced Functional Materials. 2017;**27**(41): 1703704

[25] Liu M, Johnston MB, Snaith HJ. Efficient planar heterojunction perovskite solar cells by vapour deposition. Nature. 2013;**501**(7467): 395-398

[26] Li J et al. Highly efficient thermally Co-evaporated Perovskite solar cells and Mini-modules. Joule. 2020;**4**(5): 1035-1053

[27] Deng Y et al. Tailoring solvent coordination for high-speed, room-temperature blading of perovskite photovoltaic films. Science Advances. 2019;**5**(12):eaax7537

[28] Longo G et al. Fully vacuum-processed wide band gap mixed-halide perovskite solar cells. ACS Energy Letters. 2017;**3**(1):214-219

[29] Ablekim T et al. Thin-film solar cells with 19% efficiency by thermal evaporation of CdSe and CdTe. ACS Energy Letters. 2020;**5**(3): 892-896

[30] Huang C-H et al. Deposition Technologies of high-efficiency CIGS solar cells: Development of two-step and co-evaporation processes. Crystals. 2018; **8**(7):296

[31] Roß M et al. Co-evaporated Formamidinium Lead iodide based Perovskites with 1000 h constant stability for fully textured monolithic Perovskite/silicon tandem solar cells. Advanced Energy Materials. 2021;**11**(35): 2101460

[32] Fu F et al. Monolithic perovskite-silicon tandem solar cells: From the lab to fab? Advanced Materials. 2022;**34**: 2106540

[33] Li W et al. Enabling low-temperature deposition of high-efficiency CIGS solar cells with a modified three-stage Co-evaporation process. ACS Applied Energy Materials. 2020;**3**(5):4201-4207

[34] Bonnet D. Manufacturing of CSS CdTe solar cells. Thin Solid Films. 2000; **361-362**:547-552

[35] Green MA et al. Solar cell efficiency tables (version 58). Progress in Photovoltaics: Research and Applications. 2021;**29**(7):657-667

[36] Baptista A et al. On the physical vapour deposition (PVD): Evolution of magnetron sputtering processes for industrial applications. Procedia Manufacturing. 2018;**17**:746-757

[37] Romeo A, Artegiani E. CdTe-based thin film solar cells: Past, present and future. Energies. 2021;**14**(6):1684

[38] Bonnet D, Henrichs B, Richter H. High-rate deposition of high-quality CdTe films for high-efficiency solar cells. In: The Conference Record of the Twenty-Second IEEE Photovoltaic Specialists Conference -1991. IEEE; 1991

[39] Song N et al. Epitaxial growth of Cu2ZnSnS4 thin film on Si by radio frequency magnetron sputtering. Applied Physics Letters. 2020;**116**(12): 123901

[40] Nakamura M et al. Cd-Free Cu(In, Ga)(Se,S)2 thin-film solar cell with record efficiency of 23.35%. IEEE Journal of Photovoltaics. 2019;**9**(6):1863-1867

[41] Sousa MG, da Cunha AF. Optimization of low temperature RF-magnetron sputtering of indium tin oxide films for solar cell applications. Applied Surface Science. 2019;**484**: 257-264

[42] Wang F-H et al. Effect of rapid thermal annealing time on ZnO:F thin films deposited by radio frequency magnetron sputtering for solar cell applications. Applied Physics A. 2022; **128**(3):1-8

[43] Yu J et al. Activating and optimizing evaporation-processed magnesium oxide passivating contact for silicon solar cells. Nano Energy. 2019;**62**:181-188

[44] Slauch IM et al. Model for characterization and optimization of spectrally selective structures to reduce the operating temperature and improve the energy yield of photovoltaic modules. ACS Applied Energy Materials. 2019;**2**(5):3614-3623

[45] Alberti A et al. Nanostructured TiO2 grown by low-temperature reactive sputtering for planar Perovskite solar cells. ACS Applied Energy Materials. 2019;**2**(9):6218-6229

[46] Kephart JM et al. Sputter-deposited oxides for Interface passivation of CdTe Photovoltaics. IEEE Journal of Photovoltaics. 2018;**8**(2):587-593

[47] Comparotto C et al. Chalcogenide Perovskite BaZrS3: Thin film growth by sputtering and rapid thermal processing. ACS Applied Energy Materials. 2020; **3**(3):2762-2770

[48] Arepalli VK, Shin Y, Kim J. Photovoltaic behavior of the room temperature grown RF-sputtered SnS thin films. Optical Materials. 2019;**88**: 594-600

[49] Bras P et al. Ga-grading and Solar Cell Capacitance Simulation of an industrial Cu(In,Ga)Se2 solar cell produced by an in-line vacuum, all-sputtering process. Thin Solid Films. 2017;**636**:367-374

[50] Köhler M et al. A silicon carbide-based highly transparent passivating contact for crystalline silicon solar cells approaching efficiencies of 24%. Nature Energy. 2021;**6**(5):529-537

[51] Tao K et al. The impact of indium tin oxide deposition and post annealing on the passivation property of TOPCon solar cells. Solar Energy. 2018;**176**: 241-247

[52] Parashar PK, Komarala VK. Sputter deposited sub-stochiometric MoOx thin film as hole-selective contact layer for

silicon based heterojunction devices. Thin Solid Films. 2019;**682**:76-81

[53] Truong TN et al. Deposition pressure dependent structural and optoelectronic properties of ex-situ boron-doped poly-Si/SiOx passivating contacts based on sputtered silicon. Solar Energy Materials and Solar Cells. 2020;**215**

[54] Oreski G, Stein J, Eder G, Berger K, Bruckman LS, Vedde J, et al. Designing New Materials for Photovoltaics: Opportunities for Lowering Cost and Increasing Performance through Advanced Material Innovations. Albuquerque, NM (United States): United States: Sandia National Lab. (SNL-NM); 2021

[55] Vijh A, Washington L, Parenti RC. High performance, lightweight GaAs solar cells for aerospace and mobile applications. In: 2017 IEEE 44th Photovoltaic Specialist Conference (PVSC). IEEE; 2017

Chapter 2

Sputtering Deposition

Humaira Ghazal and Nadeem Sohail

Abstract

Hundreds of research papers on various elements of sputtering have been published. The goal of this chapter is to present different aspects of sputtering that have been observed when materials are exposed to intense ion beams. Sputtering deposition is a common physical vapor deposition technology that has benefits over the molecular beam epitaxy and pulsed laser deposition in order to produce films of large area for a variety of industrial applications. Sputtering deposition has a reputation for producing high-quality epitaxial coatings and complicated oxide super-lattices at a cheaper cost than other methods, and the resulting films have proven to be essential enablers of scientific advancement. The sputtering process is discussed in detail, as well as the design and basic operations of the sputtering system, the effects of low and high energy sputtering, and changes in sputtering performance as a function of both the sputtering gas composition and the incident ion mass, dose, energy and angle. Sputtering deposition's benefits, limits, and future trends are also discussed. Sputtering deposition is an important green technology for material production.

Keywords: sputtering, deposition, physical vapor deposition, ion bombardment, film coating, emission

1. Introduction

Greene, J. E. [1] reported that the sputter deposition, described in the early 1800s, had already controlled the optical-coating industry by 1880. In 1891, radio frequency (rf) glow discharges were recorded. The term "magnetron" first appeared in literature in 1921, and the first magnetron sputtering experiments were reported in the late 1930s. In the early 1960s, capacitively-coupled rf sputtering devices were conceived and modeled. The first reactive sputtering kinetic models appeared in the 1960s. Parallel-plate magnetron was described in 1962. In 1975, a patent was submitted that led to pulsed dc and mid- frequency-ac sputtering. In the early 1980s, rotatable magnetrons were introduced. During the 1990s, two new types of magnetron sputtering emerged, both with the purpose of effectively ionizing sputter-ejected metal atoms. In 1992, tunable "unbalanced" magnetron sputtering was invented [1]. Historical flowchart is represented by **Figure 1**.

2. Sputtering deposition

Sputtering is a physical vapor deposition process (PVD) and was firstly coined by M. Blocher. This process is initiated by the bombardment of positive ions, usually Ar^+ gas

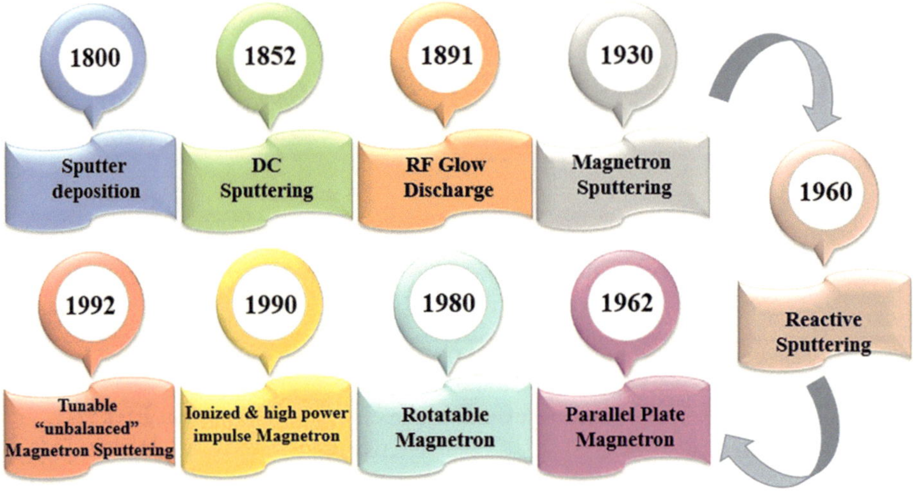

Figure 1.
Historical flowchart.

is used due to its low cost, chemical inertness and high sputtering yield provider ability. It is the process of directing high-energy ions to a target in a vacuum and remove target atoms. The deposition of these emitted atoms on the surface of substrate is called Sputter deposition.

Sputter deposition takes place in an evacuated chamber with a low pressure of a rare gas such as argon backfilled in, as shown in **Figure 2**. The film is then formed on a substrate with a dc voltage applied between a metal target (the source of the film atoms) and substrate upon which the film is deposited. The voltage causes the gas to break down into Ar + ions and electrons, forming a glow discharge. Positively charged

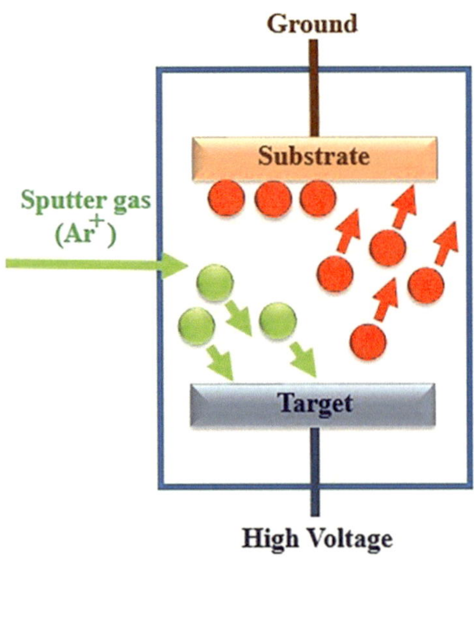

Figure 2.
Sputtering tube.

ions are accelerated and collide with the target, to sputter the target atom through momentum transfer. Some of it deposits on the substrate or sample [1].

There are some requirements for sputtering deposition as given below:

- Ion beams and potentials
  Ion beams and potentials extract ions from a plasma by means of a potential applied to the sputtering target. This sprayed plasma can be free when it occupies the entire processing volume.

- Low Pressure (less than 5 m torr)
  The pressure in the system must be low enough to prevent the sputtered particles from being subjected to excessive gas phase collisions.

- Good vacuum
  In order to maintain clean surfaces and prevent contamination from residual gas molecules, especially on the substrate, a good vacuum is required ($< 10^{-5}$ Torr).

Sputtering can also be induced by electronegative elements like oxygen and fluorine as negative ions. Reflection of high-energy ions from a sputtered surface produce high-energy neutrals [2, 3].

The energy of incident ions influences the sputtering effect. With energy less than 10 eV, the ions can also adsorb to the surface and provide that energy to phonon. At energies over around 10 keV, the ion enters the substance, passes through multiple atomic layers, transmitting the majority of its energy in the form of heat deeply into the material and altering the target materials' configuration [4].

The **Figure 3** predicts a series of collisions that occur when an ion hits the surface of a target. When the surface is bombarded with the high energy ions, emitted electrons from the metal surface are known as secondary electron. Under this ionic interaction between atoms of the substrate, ions might be neutralized or reflected.

2.1 Ionic interaction with superficial atoms

The specific methods by which atoms are ejected from a surface under ionic impact are unknown, but details of the associated interactions can be inferred. Because an ion is about the same size as an atom, when it collides with a surface, it first collides with a surface atom. The sputtering process depends on the value of the energy interchange between an incident ion and a superficial atom; it also depends on the difference in size of the incoming ion and the superficial atom. The detailed description is given below [5].

The energy exchange between a surface atom and incoming ion is substantially greater than the binding energies of the lattice atoms, hence ions impact the bombarded surface in the normal parallel direction of the surface.

As a result, the primary collision is strictly binary, with the incident particle delivering a considerable portion of its core energy to the damaged atom and keeping the rest. If the incident ion's mass is less than the mass of the surface atom it collides with, and the collision occurs in front or close, the incident ion must bounce off the surface, as seen in **Figure 4** (Event-I).

If the mass of incident ion is greater than the affected atom, both the ion and the atom will leave the collision point following inward paths from the surface, regardless of whether the collision is frontal or lateral as seen in **Figure 4** (Event-II). So we have at least

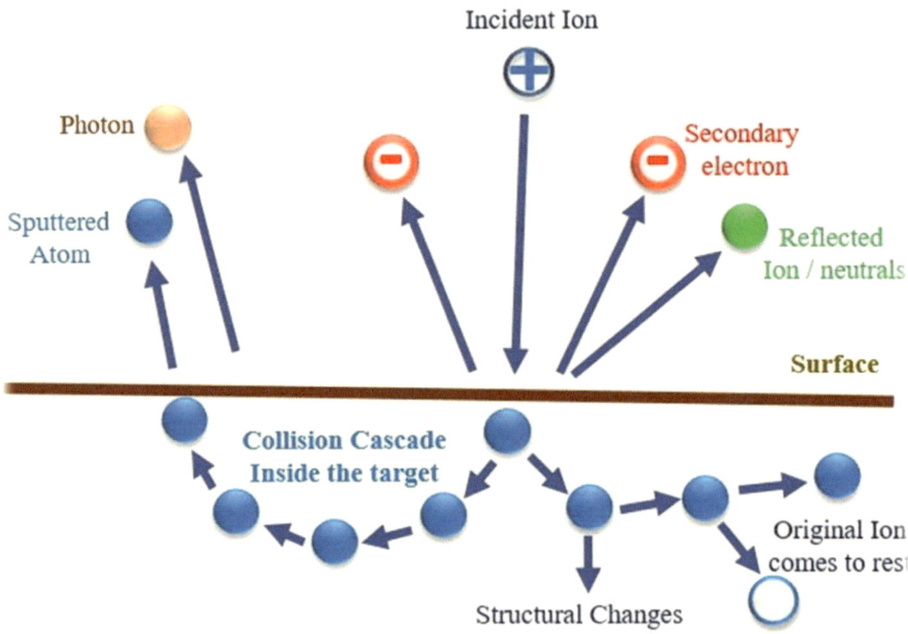

Figure 3.
Series of collisions under the ionic impact between atoms of substrate.

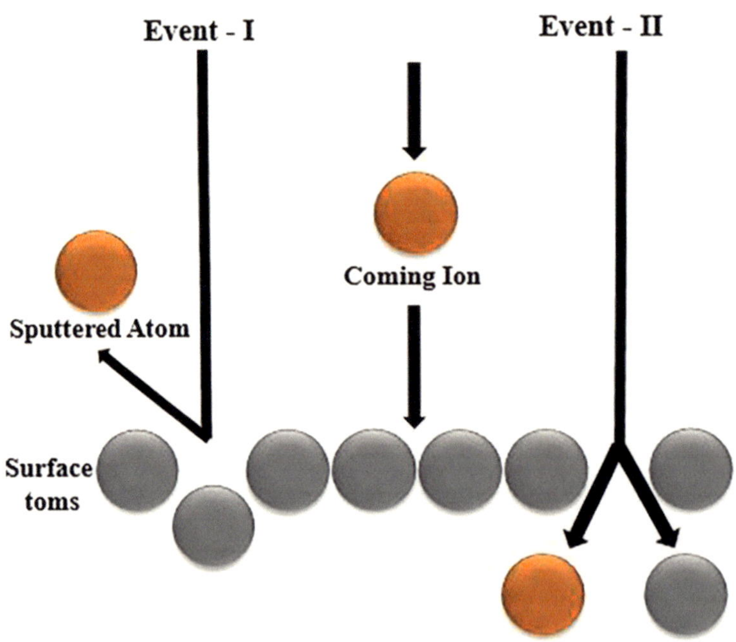

Figure 4.
Ionic interaction with superficial atoms at normal incidence.

one and usually two particles traveling to the surface with energies lower than the impacting ion's fundamental energy but still much greater than the energies of the lattice.

2.2 The impact of primary & secondary collisions on the surface

As a direct result of the primary collision, no atoms are emitted from the surface. For an atom to be emitted from the surface, it must have a velocity component in the opposite direction of the incident ion's actual velocity.

2.2.1 When the angle between ionic and damaged atomic momentum vector is 90°

As indicated by event C in **Figure 5**, the maximum feasible angle between the ion's actual momentum vector and the subsequent momentum vector of the affected/damaged atom is 90°, and the momentum vector of the hit atom is zero. In this instance, as a direct outcome of the basic collision, surface atoms cannot acquire the velocity components in the direction away from the surface. A primary collision will result in at least one and usually two secondary collisions, all of which will be close to the surface.

2.2.2 When the angle between ionic and damaged atomic momentum vector is greater than 45° (less than 90°)

We see that atoms can emerge from the surface as a direct result of the second series of binary collisions. Looking at event B in **Figure 5**, we can see that the affected ion or atom must be able to depart the collision point at an angle greater than 45°

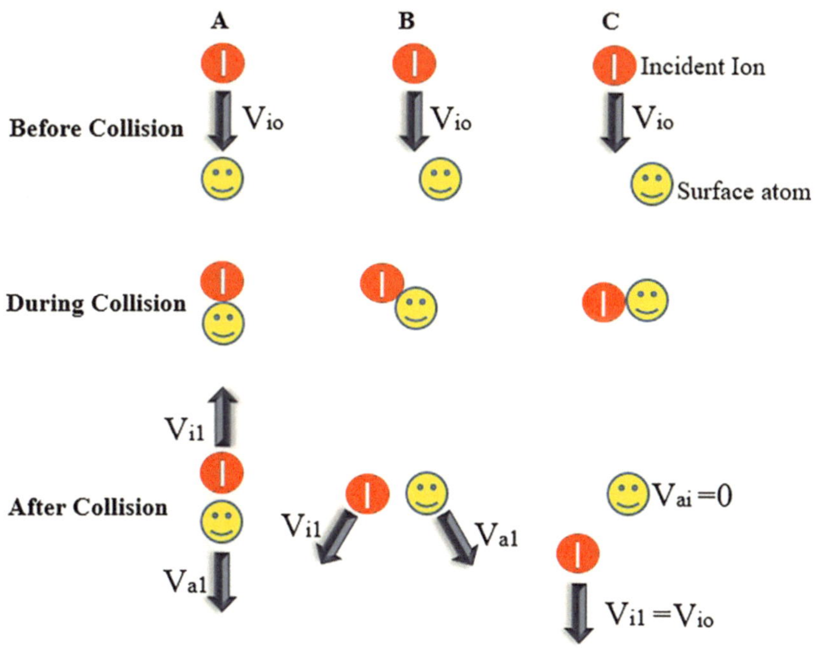

Figure 5.
Three types of collisions between ions and superficial atoms.

(relative to the actual direction of movement of the ion). As a result of the secondary impact on the same plane of motion, it should be possible for the lattice atoms to leave the secondary impact point at angles greater than 45°. Since two angles greater than 45° intersect at an angle greater than 90°, this lattice atom has an outward component of motion from the surface and can therefore be emitted.

Further consideration reveals that atoms cannot be pulled out parallel to the surface normal. Such atoms cannot be pulled in the opposite direction of the incident ion. This necessitates two 90° reflections, at least one of which moves the atom 90° in the grid/lattice while acquiring the zero velocity. Atoms with zero velocity cannot be created or ejected. Atomized atoms can be pulled towards the surface with greater force than usual, but this is not the case.

It has been determined that when ions are normally generated at the energy of interest, the sputtered atoms are removed off the surface essentially with a cosine distribution, similar to the evaporated atoms.

This is significant because the most likely emission direction is the exact opposite to the direction of the incident ion. Obviously, the incident ion's energy provided by the incident ion is so arbitrarily dispersed by the multiple collisions before the atom's emission that the incident momentum vector vanishes completely and has no effect on the emission. Note that this result applies only to the regular sputtering events [5].

2.2.3 Interaction at oblique angles

Bombarding ions can collide with the surface at oblique angles in some instances. In this instance, a basic/fundamental collision between the incident ion and the surface atom is very likely to result from a collision that occurred earlier in time. In the case of diagonal events, the incident momentum vector is found to have a significant

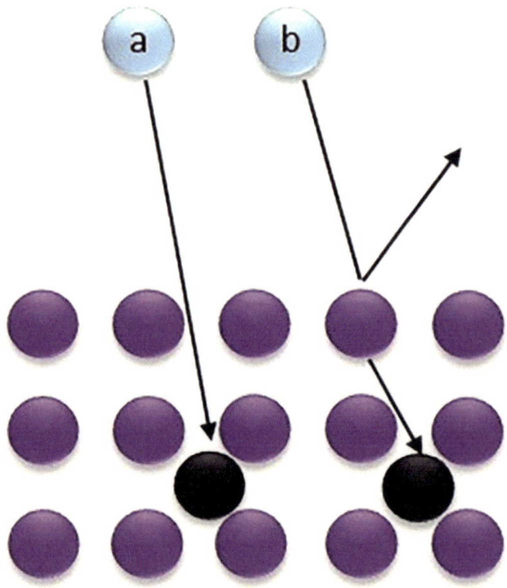

Figure 6.
(a) Direct and (b) indirect penetration by ions.

influence in the emission pattern, with the sprayed atoms being expelled extremely powerfully in the forward direction.

2.3 Direct and indirect penetration by ions

The large-angle knockout process can be broadly divided into direct and indirect as shown in **Figure 6**. "Direct" means that the surface atoms are knocked out directly by the incident ions and "indirect" means knockoff of surface atoms by incident ions just before scattering from other target atoms near the surface. Only indirect incidence works if the angle of incidence is not very oblique, but direct incidence plays a major role if the angle of incidence is grazing [6].

3. Sputtering yield

Sputter yield Y is the average number of sputtered atoms ejected by incident ions from a solid's surface per incident ion. It is represented by

$$\text{Sputtering Yield}(\text{Y}) = \frac{\textit{No.}\text{of sputtered atoms ejected from the solid}'\text{s surface } \textit{by} \text{ incident ions}}{\text{number of incident ions}} \quad (1)$$

In 1923, A. Hull discovered that the yield of sputtering was proportional to the mass of impact ions, and increased with the mass of the impact ions.

Sputtering yield increases when energy of the ion exceeds a particular threshold, regardless of substrate temperature. For smooth target faces, sputtering yield rises with increasing oblique angle bombardment up to a point, then falls as the angle of bombardment rises, resulting in increased bombarding particle reflection. The impact of changing the morphology of surface on sputtering yield have been examined by researchers. Because a large portion of the energy from high bombarding energies is deposited below the region right near the surface, energy does not directly enhance the sputtering yield.

In 1973, Attention was paid to the change in sputtering yield due to the amount of target impact on the pure metal surface. This phenomenon could be caused by bombarding species being incorporated into the region near the surface, recoil interstitials straining the near the surface region, and/or ion bombardment generating a highly defective film near the region of the surface [2].

3.1 Elements influencing the sputter yield

The interactions of incident ions with target surface atoms create sputtering. The following elements/factors will influence the sputter yield: *1) The bombarding particle's angle of incidence:* The angle of Ar atoms from the target metal's surface affects the sputter yield as well. As the incident angle increases, sputter yield rises. Between 60° and 80° angles, the sputter yield achieves the highest, the deposition rate will also be highest, resulting in a thicker coating. Further increasing the angle will cause the sputtering yield and film thickness to decrease rapidly. *2) Sputtering Voltage through which the ion is accelerated:* The applied voltage regulates the maximum energy that the atoms that are expelled from the target can have. Ions' Kinetic energy (KE) that collide with the target surface is controlled by the applied voltage. The energy of the ions will be greater due to the higher cathode voltage, which will cause a greater number of atoms to sputter from the target. The more atoms that are sputtered, the more material will be deposited and sputter yield will rise as a result. The term "Threshold Voltage" refers to the required minimum voltage for the sputtering process. Ions do not have enough energy to knock out the target's binding energy atoms below this voltage. Its value ranges from 0 to 100 eV. To have an appropriate film thickness, the normal voltage range is between 100 and 1000 V. Unfortunately, raising power or voltage has a lot of negative consequences such as any energy used to operate the gun will eventually be lost, about 75% of it goes up heating the cooling water for the gun. Thermal conductivity, melting point, thermal coefficient of expansion and mechanical strength properties of the target are undoubtedly important factors. *3) Sputter Gas Pressure:* The mean free path will be shorter and there will be more collisions before the sputtered atoms deposit on the substrate as the gas pressure rises. A low deposition rate brought on by increasing number of collisions will result in a reduction in film thickness. As a result, modest/low pressures between 10^{-5} and 10 torr are used during the sputtering process. A small increase in deposition rate is produced by lowering the sputter gas pressure through two mechanisms: I- There will be fewer thermal collisions for sputtered atoms that are leaving the target. They are more likely to propagate to the substrate, less likely to scatter laterally and enhancing the deposition rates slightly. II- The plasma-to-target voltage will slightly rise in power control mode when using RF or DC power. As a result, the energy of the ions that collide with the target will be higher, somewhat increasing the sputter. A change in film homogeneity is one potential adverse effect of lowering the gas pressure. Deterioration is usually unpredictable because many factors play a role. The quantity of thermalizing impacts is lessened, though, and this is an obvious aspect. Arcs are more likely to form close to the target as a result of the combination of lower gas pressure or higher plasma to target voltage. According to the experimental results described by Chargui A, et al. [7], higher pressure results in thinner and less crystallinity in the tungsten films produced. Tungsten foil exhibit the best electrical and elastic properties at low pressures.

4. *Increasing Target Size:* The sputter rate increases with target diameter. This can be explained easily. For a given power density, a bigger target diameter results in a larger sputter trench area, and a larger trench area results in a higher sputter rate. 5. *Number of Guns:* Most R and D deposition systems are equipped with multiple sputter guns. The user often installs several target materials in each pistol. The sputter rate and subsequent deposition rate can be doubled, tripled, etc. when the identical target material is added into two or more guns and they are all fired at the same time. The disadvantage is that many multi-gun systems only have one power source and were not designed for simultaneous deposition operations. This method could be more expensive if additional supplies are needed for concurrent operation. 6. *Atomic number of element:* By reducing the target element's atomic number, the spatter yield is increased. 7. *Reducing the target-to-substrate distance:* A quick, easy technique to boost deposition rate is to shorten the throw distance—the distance between the target and the substrate. The flux distribution of a sputtered material is described by terminology like over-cosine and under-cosine. Material is ejected from a circular 'trench' around the target. For these remarks, the arrival rate of the sputtered particles (per unit area of the substrate) varies as the inverse square of the throw distance. This means that halving the throw distance will quadruple the rate at which the material arrives at the substrate and the layer thickness will be four times the previous rate.

It is crucial to take into account how the shorter throw distance would affect the uniformity of the film's (thickness). The number of thermalizing collisions between sputtered atoms and sputter gas atoms increases with throw distance, for example, if material departs the target in an approximately cosine distribution pattern. The cosine distribution tends to "flatten out" as a result of these encounters, which makes the deposition more uniform across the substrate. Film homogeneity may be worse at shorter distances because there are fewer impacts at shorter throw distances. Additionally, substrates may experience greater energy sputter particles, more stray electrons, more plasma ions and "hot" neutrals, as well as increased thermal radiation heat transfer from the plasma and target surface, when throw distances are shorter. Excessive outgassing of the substrate, an increased compressive stress of the growing membrane/film, substrate melting, substrate films/ membranes under the film destroyed by electron bombardment and other negative effects are caused by shorter throw distances. There are also advantageous effects of shorter throw distances (higher substrate temperatures) such as tensile stress of film may be lowered, the high energy of the incoming atoms improves the adhesion of the film and the membrane can be "densified" by colliding high-energy plasma ions with "high temperature" neutral particles. *8. Temperature:* The system's temperature has an impact on the thickness of the film as well. The atoms on the surface will be more mobile as the temperature rises. The larger particle size and smoother film will result from this higher mobility. The larger the particle size, the thicker the film and greater the sputter rate. The substrate is often placed on a heating stage for the higher temperature because of this. However, the act of sputtering itself generates heat as a result of collision between the atom and the surface. The deposited film might be harmed by excessive heating. As a result, cooling is required in cases of severe heat [8–10].

3.2 Cosine sputtering law

The angular distribution of the sputtered particles ejected from the target surface can often be estimated by a cosine distribution as shown in **Figure 7** in circumstances of normal incidence of the projectile atoms on the surface of the target.

$$J\Omega(\theta) = Y\Phi\left[\frac{cos\theta}{\pi}\right] \tag{2}$$

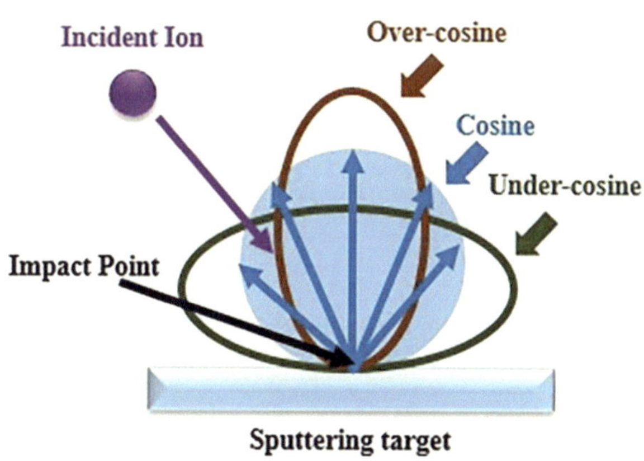

Figure 7.
Cosine law angular distribution.

Here, $j_{\Omega}\phi$ is the angular distribution of the emission flux as a function of the angle Θ into the differential solid angle $d_{\Omega}(\phi)$, Y is the sputter yield, and ϕ is the local ion flux incident on the surface.

If the recoil velocities of the sputtered atoms are considered to be isotropic, this conclusion can be calculated analytically. Low sputter-ion energies (undercosine) and high ion energies (overcosine) show deviations from the ideal cosine distribution. Undercosine means the flatter distribution and overcosine means the sharp forward-peak distribution [11].

4. Types of sputtering deposition

There are many sputtering deposition processes like Gas flow sputtering/Glow discharge sputtering deposition, Ion beam sputter deposition (IBSD), Reactive sputter deposition, Ion-assisted deposition, Magneton sputter deposition and Radio frequency (RF) sputter deposition etc.

Few mechanisms of sputtering deposition are explained in detail.

4.1 Ion beam sputtering deposition (IBSD)

Ion beam sputter deposition (IBSD) can solve a variety of problems. IBSD, unlike other PVD processes, provides a unique desirable ability to modify properties of thin film such as dense film, fewer flaws, higher purity, greater adhesion etc.

A setup is depicted in the **Figure 8**. An ion beam source, target, and substrate holder make up the ion beam sputter deposition setup. To sputter a target, IBSD uses a wide beam ion source with low energy ions. A film forms when the powder (emitted particles) condenses on a sample. Primary particles also disperse on target to provide aid in the formation of a thin coating. Dispersed primary particles as well as the sputtered target particles, both play an important role in the film formation process. The essential process parameters are geometric parameters and ion beam parameters, such

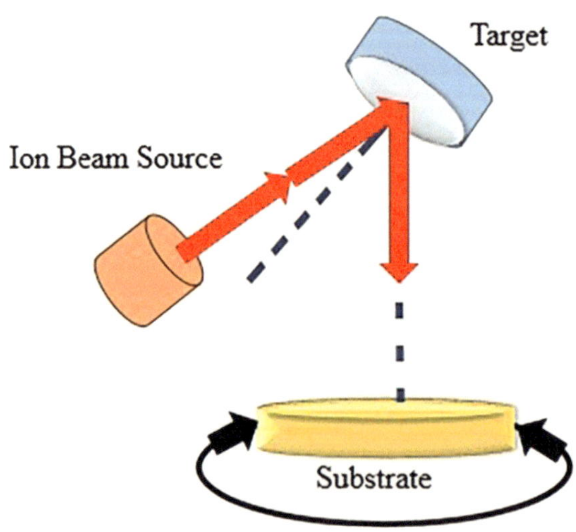

Figure 8.
Ion beam sputtering.

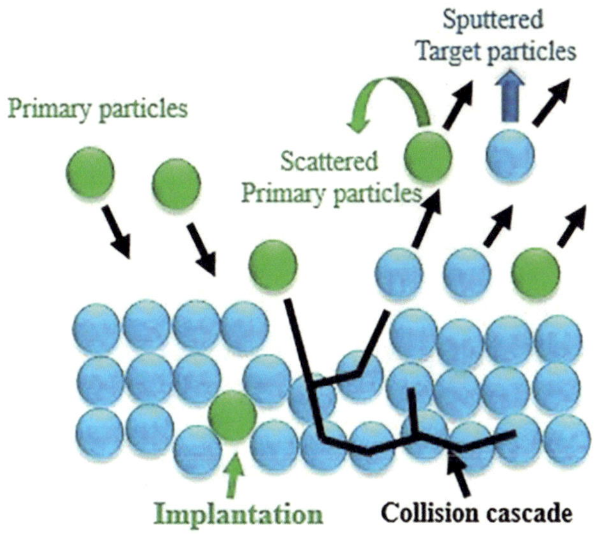

Figure 9.
Schematic diagram of ion surface interaction and resulting implantation, scattering and sputtering process.

as ion's angle of incidence, polar emission angle, angle of scattering and ion energy E_{Ion} etc. When these parameters are changed, energy distributions of the particles that make up film are also changed.

When an energetic particle collides with a target, the momentum and energy are transmitted from primary to the target particle. Sputtering, dispersion, and implantation are all significant processes in IBSD as in **Figure 9**. The retreating target particles and the scattered primary particles can collide further, creating a collision cascade or leaving the target. If target particles at the surface have acquired enough energy to overcome the binding energy of the surface, they can escape the target. The dispersed primary particles can either scatter or stay on the target. The dispersed particles are known as scattered particles.

Two particle collision is shown in **Figure 10**. Here, α is the Ion Incident angle, β is the Polar emission angle and γ is Scattering angle γ. Angle of incident plays an important role to decide either the resulting process will be scattering or sputtering.

Additional features must be addressed in order to fully utilize IBSD's potential. To begin with, the ion beam is always slightly divergent. As a result, the primary particles will collide with the chamber's components and walls. The forming film can be contaminated by eroded particles, so the chamber size must be large enough to reduce or avoid this. Second, the vacuum system's pumping speed must be high enough to prevent the background gas particle coverage of the target surface. The background pressure, current density of ion beam, and target body all play a role in surface coverage.

Ion beam sputtering deposition (IBSD) disadvantages include a slower growth and more difficult scaling. In addition, ion beam sources are extra complex than magnetrons, incorporating peripheral components [12].

4.2 Magnetron sputtering deposition

In the Magnetron sputtering technique, permanent magnets are used. As illustrated in the **Figure 11**, these magnets are placed behind the target to generate a

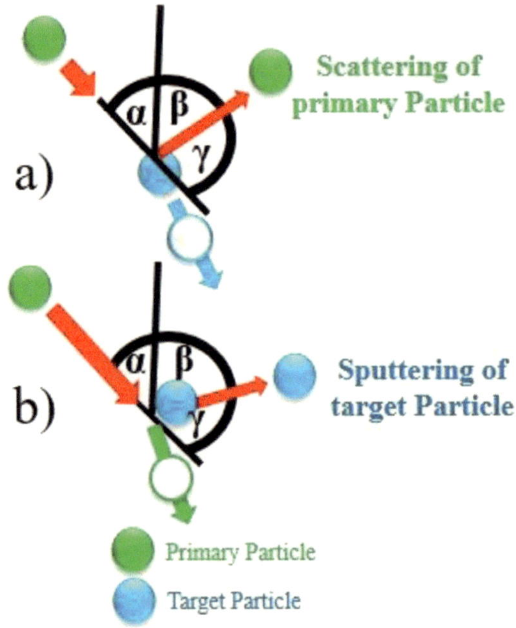

Figure 10.
Two particle collision (a) direct scattering and (b) direct sputtering processes.

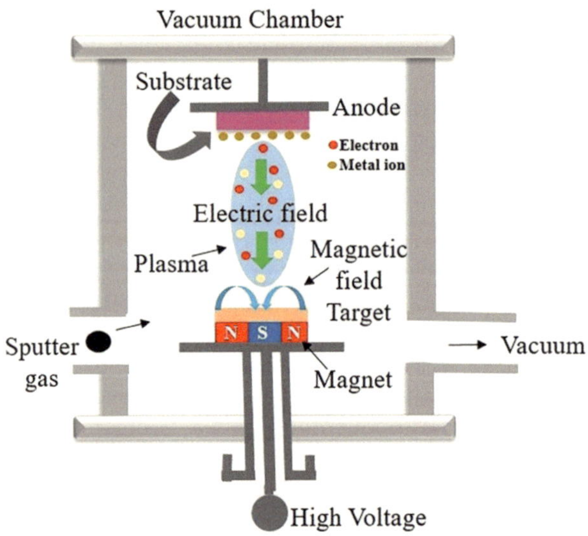

Figure 11.
Magnetron sputter deposition.

magnetic field. As ions are heavier than electrons so ions are scarcely influenced directly by the magnetic field. But the magnetic field causes the electrons to flow on a spiral course, extending their residence duration in the plasma. Now the likelihood of electrons colliding with background gas atoms will raise, causing a significant number of gas atoms to ionize.

In the presence of an electric field, gas ions speed up the bombardment of the target to produce sputtering atoms which eventually condense on the substrate to produce the required film. We conclude that the ionization rate and ion bombardment rate at the target surface raise under the effect of an electric field, and hence sputter rates increase [13].

4.2.1 Magnetron configuration

There are three main kinds of magnetron configuration as illustrated below.

1. Conventional magnetron/Balanced magnetron

2. Unbalanced magnetron

3. Closed-field unbalanced magnetron sputtering (CFUBMS)

4.2.2 Conventional magnetron/balanced magnetron

In 1986, papers to describe various magnetic field configurations of the substrates were published by Window and Savvides [14] which gave birth to the terms "balanced" and "unbalanced" magnetrons. Shortly thereafter, unbalanced magnetron technology was used to deposit thin films.

In a balanced magnetron (**Figure 12**), all lines of magnetic force originate from one pole and are closed by the other pole of the target. The magnetic flux strengths via the pole faces of the outer and inner magnets are equal or comparable. As the magnetic field controls the plasma transmission path, the balanced magnetron's plasma is restricted or trapped at the cathode, which is advantageous for high rate cathode sputtering, but resulting production efficiency is relatively low and also prevents the large amount of bias current from reaching the substrate [15].

4.2.3 Unbalanced magnetron

The balanced magnetron can be unbalanced if the outer or inner set of magnets is strengthened or weakened. As the ion bombardment of growing films supports in

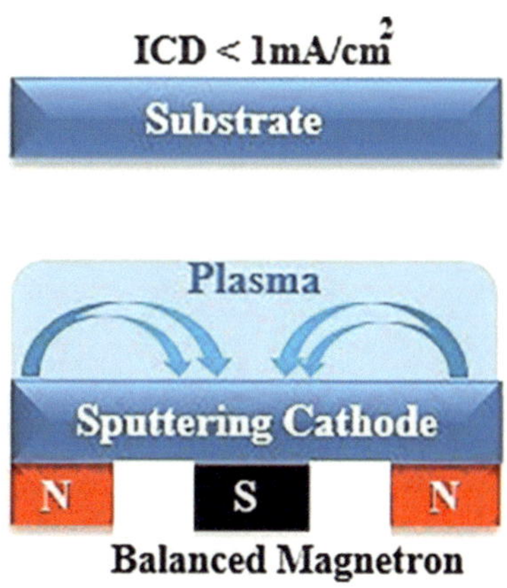

Figure 12.
Balanced magnetron.

the formation of thin films with better characteristics. The technology of unbalanced magnetron sputtering allows to get thin films with such improved properties. Fraser and Cook [16] initially highlighted the possibilities of this technology in 1977. It's been utilized to deposition and ion-bombard films at the same time.

By allowing some of the confined lines of magnetic field, which are parallel to the surface of cathode in balanced magnetron, to become perpendicular to the cathode surface in the unbalanced magnetron, the plasma purposely leaks and collide with the substrate. As a result, some of the electrons discharged from the cathode surface are allowed to exit the cathode region along those normal field lines, causing electrostatic forces to pull ions along with them. This produces a plasma beam that is focused towards the substrate surface, bombarding it as well as the forming film surfaces with ions from the sputtering gas.

As the mobility of electrons is significantly more than ions and their mean free path is also longer, so the surface of an insulating or isolated substrate submerged in this high density plasma will attain a negative charge and potential with respect to the plasma until the electron and ion fluxes become equal. The ions will collide with substrate under the effect of this floating potential [15].

4.2.4 Types of unbalanced magnetron

Savvides and Window [17] also introduced types of unbalanced magnetron, type I and type II as in **Figure 13**. Magnetic flux is important in both type I and type II.

The inner set magnetic flux in type I is greater than the outer set magnetic flux, whereas the inner set magnetic flux in type II is lower than the outer set magnetic flux. To put it another way, in type I, all field lines originate from the inner set of

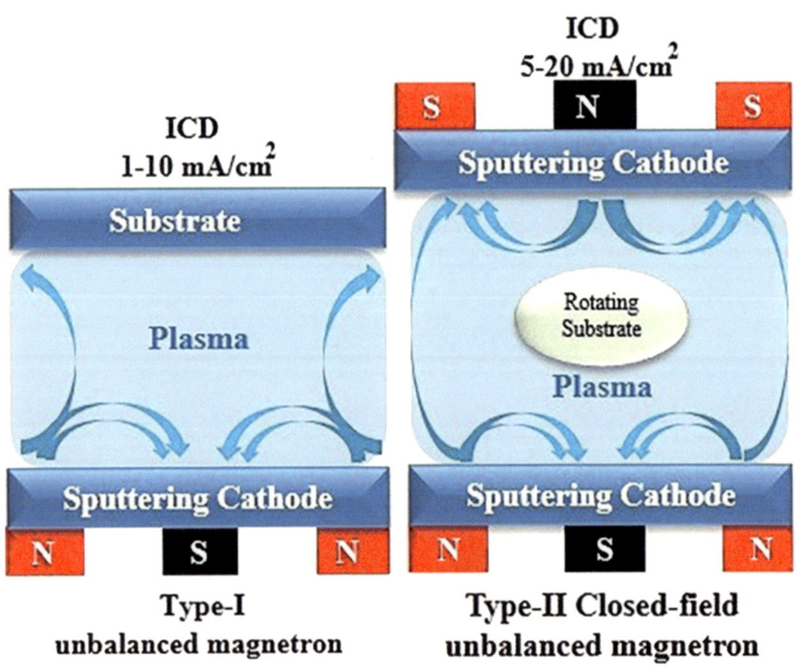

Figure 13.
Types of unbalanced magnetron.

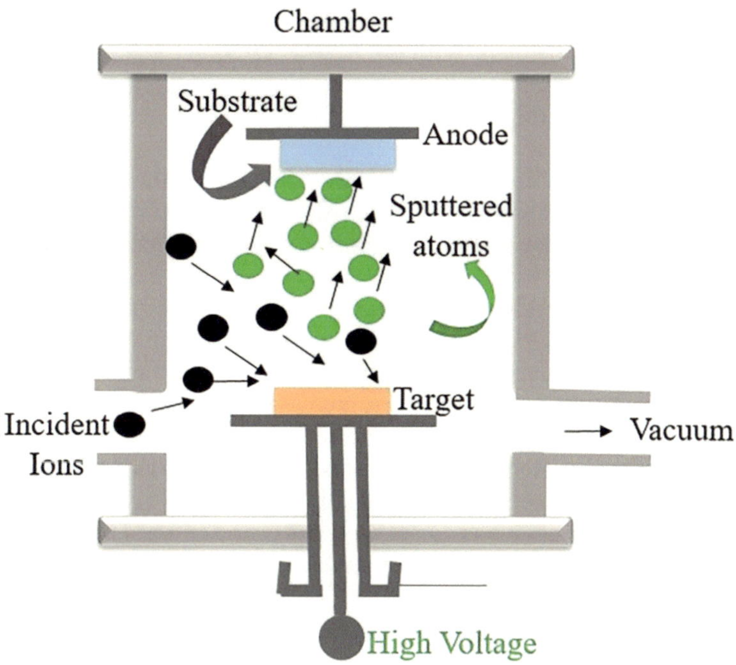

Figure 14.
DC sputter deposition.

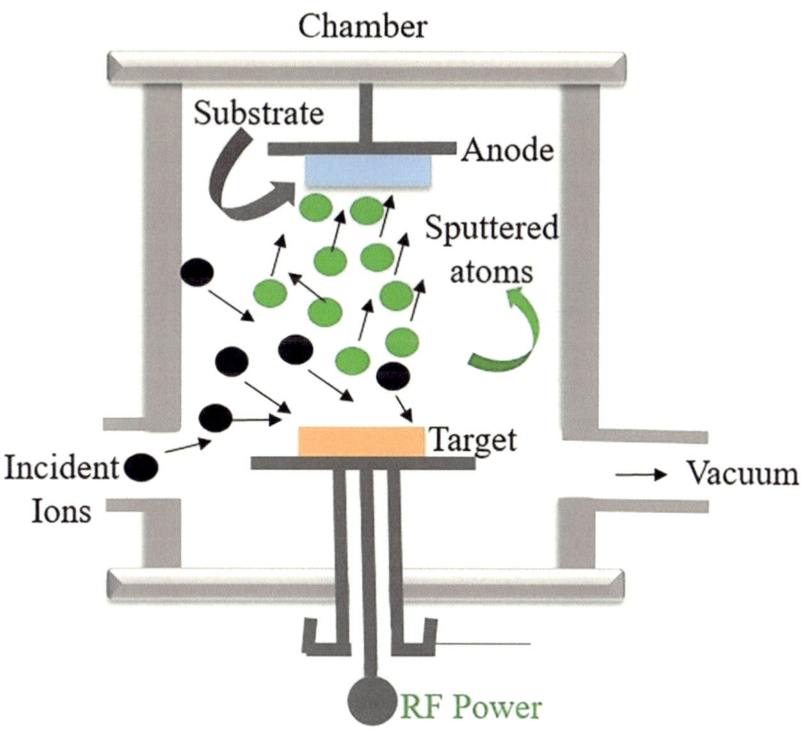

Figure 15.
RF sputter deposition.

magnets, with only a few reaching the outer set, whereas in type II, field lines comes out from the outer set of magnets, with only a few reaching the inner set.

4.3 DC and RF sputtering deposition

The power source in DC sputtering is direct current. Positively charged sputtering gas is propelled towards the target in this approach. As a result, atoms are ejected and deposited on the surface of substrate. Schematic diagram of DC is shown in **Figure 14**.

A cathode (the target) and an anode are connected in series with the blocking capacitor in RF sputtering as shown in **Figure 15**. The capacitor detects that power from the RF source is transferred to a plasma discharge. There are two stages of RF sputtering. The target material is negatively charged in the first cycle. Atoms get polarized as a result of this. The atoms of sputtering gas are drawn to the source that knock away source atoms. Here, the polarization of the target leaves source atoms and ionized gas ions on the surface of the target. The target is positively charged in the second cycle. This causes gas ions and source atoms to be emitted by depolarization. These are propelled towards the substrate, causing deposition [18].

5. Application of sputtering deposition processes

There are many advantages of sputtering deposition, some are listed below

- Deposition of any solid is possible.
- The vaporization source is provided by the sputtering target, which is steady and long-lasting.
- Vaporization can take place from all sides of a solid surface.
- Sputtering the target in specific ways results in a large area vaporization source.
- The sputtering target can give specialized vaporization geometries in some setups.
- Surface preparation on-site is simple to add into the process.
- Sputtering targets can be adapted to the surface of a substrate such as a cone or sphere.
- It is a flexible method for producing specialized nanostructures, stable colloidal particles, metallic, semiconductor and magnetic nanoparticles on the liquid surfaces.
- Films with ion assisted deposition can be used to make films for engineering applications, an appealing end to a product, and such ornamental films soak up a big part of the industry. For example, sanitary objects with a silver or gold finish contribute to the product's appeal [2, 19, 20].

6. Conclusion

Sputter deposition was long regarded a black art, but today's spectrum of material types that can be deposited has greatly expanded. Sputtering technologies with low cost, the capacity for large-area films, and unique kinetic regime of deposition material can be used to create epitaxial complex oxide, carefully regulated heterostructures, superlattices of the best quality, and film repeatability.

Sputtering yield is affected by a variety of factors, including sputtering angle, incidence ion mass, dosage, and energy of impact ions. If the sputtering pressure is too high, material dispersion back to the target reduces the deposition rate dramatically. Higher pressures alter the film growth process and film morphology significantly. High Vacuum conditions are required for accurate measurements of the absolute sputtering yield so that impurities do not deposit on the sample during analysis. When it comes to designing thin film characteristics, the reactive sputtering technique is crucial. IBSD is a flexible method for customizing the properties of film-forming particles. Due to its flexibility and scalability, magnetron sputtering remains an intriguing and widely utilized technology, and has thus established a strong position for large-area thin film deposition. This explains why sputter deposition has such a broad range of applications. Sputtering procedures that lower impurity content are now available. As a result, research into a fuller description of this technique is ongoing.

This is a strong technology for creating materials that cannot be made any other way, and it'll only get more essential as time goes on. Despite how simple the approach is to apply; it poses enough hurdles to be scientifically fascinating. This explains why it is so popular among academics. There are, nevertheless, some unexpected elements to be uncovered, explained, and used.

Acknowledgements

Big Thanks to our beloved Allah Almighty. Author Humaira Ghazal would also like to thank honorable parents (loving mother and great father Muhammad Yousaf Ishaq) and respected teacher Dr. Muhammad Rizwan Riaz (UET Lahore) who always encourage and support her in her research work.

Conflict of interest

The authors declare no conflict of interest.

Author details

Humaira Ghazal[1*] and Nadeem Sohail[2]

1 University of the Punjab, Lahore, Pakistan

2 University of Management and Technology Sialkot Campus, Lahore, Pakistan

*Address all correspondence to: 3520226945786@pu.edu.pk

References

[1] Greene JE. Review article: Tracing the recorded history of thin film sputter deposition: From the 1800s to 2017. Journal of Vacuum Science & Technology A: Vacuum, Surfaces, and Films. 2017;**35**(5). DOI: 10.1116/1.4998940

[2] Mattox DM. Physical sputtering and sputter deposition. The Foundations of Vacuum Coating Technology. 2nd ed. 2018. pp. 87-149. DOI: 10.1016/b978-0-12-813084-1.00004-2

[3] Mattox DM. Physical sputtering and sputter deposition (sputtering). In: Andrew W editor. Handbook of Physical Vapor Deposition (PVD) Processing. 2nd ed. Oxford, UK, Burlington, MA, USA: Elsevier; 2010. pp. 237-286. DOI: 10.1016/b978-0-8155-2037-5.00007-1

[4] Guest. Module 13 sputtering class notes - Chemical engineering. n.d. Available from: moam.info.https://moam.info/module-13-sputtering-class-notes-chemical-engineering_59f54d4e1723dd91f52db1a3.html

[5] Stuart RV. Vacuum technology, thin films, and sputtering: An introduction: 9780123960542: Books - Amazon.ca. n.d. Available from: https://www.amazon.ca/Vacuum-Technology-Thin-Films-Sputtering/dp/0123960541. [Accessed: July 14, 2012]

[6] Delpha D, Mahieu S. Reactive sputter deposition, Springer Series in Materials Science. 2008. Available from: https://link.springer.com/book/10.1007/978-3-540-76664-3?noAccess=true

[7] Chargui A et al. Influence of thickness and sputtering pressure on electrical resistivity and elastic wave propagation in oriented columnar tungsten thin films. Nanomaterials. 2020;**10**(1):81

[8] Ejaz H, Hussain S, Zahra M, Saharan QM, Ashiq S. Several sputtering parameters affecting thin film deposition. Journal of Applied Chemical Science International. 2022;**13**(3):41-49

[9] Kurt J, LESKER company. Practical process tips. 2010;**7**(1). Available from: https://www.lesker.com/leskertech/archives/0g11m3h/leskertech_v7_i1.pdf

[10] Shehataa MM, Radwanc SI, Reheemc AA. SRIM simulation of parameters affecting sputtering yield and chemical structure using Kr, Cu and Pb ions bombardment. Arab Journal of Nuclear Sciences and Applications. 2017;**50**(4):193-200

[11] Seshan K. Handbook of Thin Film Deposition. William Andrew. Google. (n.d.). Handbook of Thin Film Deposition. Google Books; 2012

[12] Bundesmann C, Neumann H. Tutorial: The systematics of ion beam sputtering for deposition of thin films with tailored properties. Journal of Applied Physics. 2018;**124**(23):231102. DOI: 10.1063/1.5054046

[13] Deng Y, Chen W, Li B, Wang C, Kuang T, Li Y. Physical vapor deposition technology for coated cutting tools: A review. Ceramics International. 2020;**46**(11):18373-18390. DOI: 10.1016/j.ceramint.2020.04.168

[14] Window B, Savvides N. Charged particle fluxes from planar magnetron sputtering sources. Journal of Vacuum Science and Technology A. Vacuum, Surfaces, and Films. 1986;**4**(2):196-202

[15] Safi I. A study of reactive magnetron sputtering of alloy transparent conducting oxides from elemental targets

[Doctoral dissertation]. Loughborough University. 1997

[16] Fraser DB, Cook HD. Film deposition with the Sputter Gun. Journal of Vacuum Science and Technology. 1977;**14**(1):147-151

[17] Savvides N, Window B. Unbalanced magnetron ion-assisted deposition and property modification of thin films. Journal of Vacuum Science and Technology A. 1986;**4**(3):504-508

[18] Home IFN Trento. RF-sputtering principles. (n.d.). 2022. Available from: https://www.tn.ifn.cnr.it/facilities/rf-sputtering-facility/rf-sputtering-principles

[19] Wender H, Migowski P, Feil AF, Teixeira SR, Dupont J. Sputtering deposition of nanoparticles onto liquid substrates: Recent advances and future trends. Coordination Chemistry Reviews. 2013;**257**(17-18):2468-2483. DOI: 10.1016/j.ccr.2013.01.013

[20] Colligon JS. Ion-assisted sputter deposition. Philosophical Transactions of the Royal Society of London Series A: Mathematical, Physical and Engineering Sciences. 2004;**362**(1814):103-116. DOI: 10.1098/rsta.2003.1303

Chapter 3

Thin Films Processed by SILAR Method

Md Abdul Majed Patwary

Abstract

SILAR is one of the simplest techniques in terms of the better flexibility of the substrate choice, capability of large-area fabrication, deposition of stable and adherent film, low processing temperature for the film fabrication as well as reproducibility. This technique is very budget friendly since it does not require any sophisticated equipment. Moreover, various fabrication parameters such as solution concentration, precursors, the number of cycles during immersion, pH, annealing, doping, and growth temperature affect the rate of fabrication as well as the structural, optical, and electrical properties of the fabricated thin films led the technique unique to study in an extensive manner. A chapter regarding different aspects of semiconductors-based optoelectronics by SILAR has yet to be published. This chapter will concern the recent progress that has recently been made in different aspects of materials processed by the SILAR. It will describe the theory, mechanism, and factors affecting SILAR deposition as well as recent advancements in the field. Finally, conclusions and perspectives concerning the use of materials in optoelectronic devices will be represented.

Keywords: SILAR, thin films, supercapacitors, photovoltaics, water splitting

1. Introduction

The successive ionic layer adsorption and reaction (SILAR) technique was first introduced in 1985 by Nicolau for the deposition of ZnS and CdS [1], and Ristov et al. for the deposition of Cu_2O thin films [2]. Nicolau applied the adsorption technique of film preparation, relating two sources—a metal source of aqueous solutions of $CdSO_4$ or $ZnSO_4$, and a sulfide source of Na_2S, maintaining ambient temperature. The substrate was dipped in each of the solutions following consecutive cycles, applying rinsing steps in between to eliminate excessive precursors. In the case of Ristov et al., the method was employed using clean glass substrates, which were sequentially immersed into a solution of NaOH and a Cu-complex such as $[Cu(S_2O_3)]^-$ to prepare Cu_2O thin films. The complex $[Cu(S_2O_3)]^-$ was well-maintained at ambient temperature, while the NaOH solution was kept at a temperature range from 60 to 80°C. The outstanding structural, electronic, and optical properties of the thin films produced in the two investigations stimulated the persistence of the practices. Hence, the technique was functional to the fabrication of a large variety of thin film semiconductors till to date [3–5]. Consequently, SILAR appeals huge scientific attention and has been well known as a facile technique for the fabrication of thin films of metal oxides, sulfides, selenides, peroxides, and hydroxides, as well as more complex hetero-structured thin films [6–11]. Compared with the other popular deposition techniques, SILAR is unique due to multiple reasons [3, 12–14] as mentioned below:

i. The thin film is generally fabricated on a simply cleaned and rigid planar substrate having no dimensional limits. There is no restriction on the surface of the coating materials, so films can be fabricated on temperature-sensitive substrates like plastics. Even oxidation or corrosion of metal backing substrates can be used to deposit films by choosing suitable precursors.

ii. The deposition rates and film thickness can be well controlled by monitoring reactant precursors, which are generally the desired cationic and anionic salts dissolved in solvents, while the anticipated stoichiometry can be attained through changing their type, concentration, or other involved dipping parameters.

iii. By controlling the number of deposition cycles and concentration of species, the thicknesses of the thin films can be simply tuned over a wide range such as from nm scale to μm.

iv. SILAR fabrications are very convenient and energy efficient as the technique is mostly controlled at room or low temperature. If required, the as-deposited coating materials can be annealed post-deposition to activate grain growth, crystallization, etc.

v. Besides, the fabricated thin films can be reformed to show preferential crystallographic orientation as well as grain assembly due to the controllability of ionic reactions performed at the substrate solution interface.

vi. The SILAR method supports film development only on the surface of the substrate that is immersed into the solution, hence diminishing unnecessary consumption of the used reactants. If required, precursor solutions could also be reloaded and reprocessed.

Therefore, SILAR is a vastly multipurpose and influential process for the fabrication of numerous thin film materials having huge technological attention and, hence, unlocked a wide window in optoelectronic device applications.

2. Theory and process mechanism

SILAR is widely used, simple technique to fabricate high-quality thin films [3, 15]. During deposition, successive ionic layer adsorption and reaction of the ions take place at the solid-solution interface of the substrate. Thus, the thin film of the compound, A_pB_y is deposited on to the substrate surface by dint of the adsorbed cations, xA^{y+} and anions, qB^{p-} due to the following heterogeneous chemical reactions:

$$A_xQ_y\ (s) \rightarrow x\ A^{y+}\ (aq) + y\ Q^{x-}\ (aq) \qquad (1)$$

$$P_pB_q\ (s) \rightarrow p\ P^{q+}\ (aq) + q\ B^{p-}\ (aq) \qquad (2)$$

$$A^{y+}\ (aq) + B^{p-}\ (aq) \rightarrow A_p\ B_y\ (s) \downarrow \qquad (3)$$

where x, y, p, q and y^+, q^+, x^-, p^- are the number and charges of the corresponding ions A (metal ions), P (cationic precursor), Q (anionic precursor), and B (anions) respectively [2, 16]. Sometimes, the ligands L_n are a necessity to complete the reaction [17–20]. The solution having the first element containing the final target material can be thought as the compound A_xQ_y fully dissociated in the chosen solvent such as in water (Reaction 3). Usually, A_xQ_y is a metal salt where A^{y+} represents cations such as Zn^{2+}, Cu^{2+}, Mn^{2+}, Cd^{2+}, Bi^{3+}, and B^{p-} represents anions such as NO^{3-}, Cl^-, SO_4^{2-}.

Hence, a basic SILAR cycle comprises four different steps, correlating alternate immersion of the substrate into cationic and anionic precursor solution followed by rinsing in every immersion cycle to eliminate loosely adhered particles as shown in **Figure 1** and described below:

2.1 Adsorption

First step of the SILAR process is the formation of the Helmholtz double layer, which is due to the initial adsorption of cationic precursor, xA^{y+}, on the surface of the substrate. This layer is generally composed of two charged layers, the positively charged inner layer and negatively charged outer layers. The positive (+ve) layer consists of the cations, xA^{y+}, while the negative (−ve) layer, yQ^{x-}, is the counter ions of the cations.

2.2 Rinsing I

In the second step, excessive adsorbed ions, xA^{y+} and yQ^{x-}, are rinsed away from the diffusion layer toward the bulk solution and a hypothetical monolayer is formed. This results in a saturated electrical double layer showing an ideal scenario of the process.

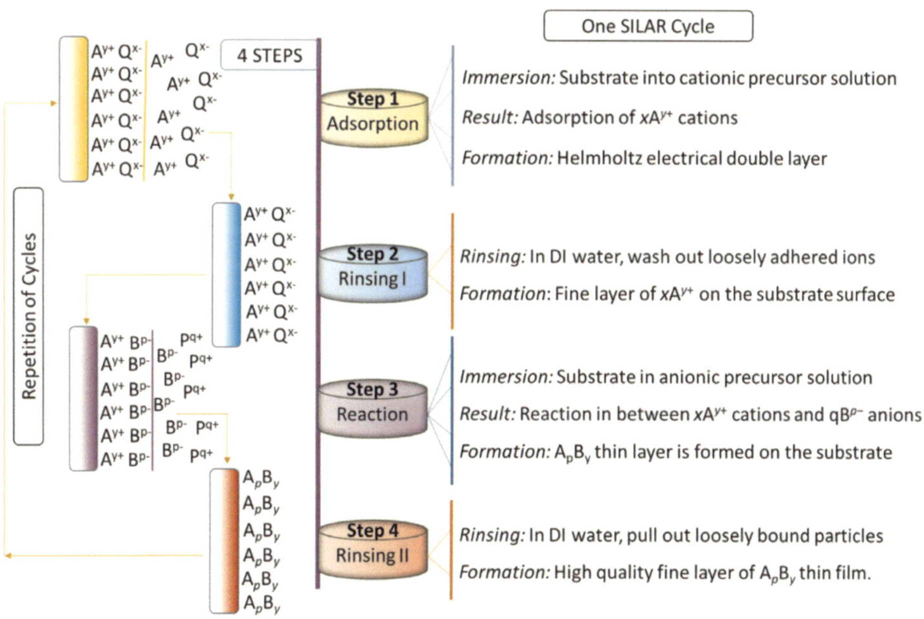

Figure 1.
Representation of different steps during a SILAR cycle.

2.3 Reaction

In the reaction stage, the anions, qB^{p-}, from anionic precursor solution are introduced into the system. A solid substance, A_pB_y, is formed on the interface due to the low stability of the material. This process employs the reaction of xA^{y+} surface species with the anionic precursor, qB^{p-}.

2.4 Rinsing II

In the final step of a SILAR cycle, the excess and unreacted species (yQ^{x-}, pP^{q+}) and the reaction by product from the diffusion layer are removed leaving expected films.

A schematic presentation of a single cycle for the fabrication of Cu_2SnS_3 film is shown in **Figure 2** [21]. In the case of Cu_2SnS_3 film fabrication, ion-by-ion type of deposition takes place through nucleation spots of the adsorbed surfaces [22]. Nucleation occurs due to the surface condensation of the ions and outcomes of, that is, an dense adherent thin film [23]. The substrate was firstly dipped into the cationic precursor containing mixed $CuCl_2$ and $SnCl_2$ solutions, where Cu^{2+} and Sn^{2+} species were available. Sn^{2+} ion in solution is good reducing agents, and thus, Cu^{2+} reduces to Cu^+ and Sn^{2+} is oxidized to Sn^{4+} in cationic solution as shown by the following reaction:

$$2CuCl_2 + SnCl_2 + 6H_2O = 2Cu^+ + Sn^{4+} + 6HCl + 6OH^- \tag{4}$$

The substrate was then rinsed off with DI H_2O to eliminate the loosely bounded reactants. Then, it was dipped into an anionic precursor containing $Na_2S.xH_2O$ solution, which gave sulfide ions (S^{2-}) to react with the cations Cu^+ and Sn^{4+}. Finally, the reaction occurred between the pre-adsorbed Cu^+, Sn^{4+} cations, and the S^{2-}anion to form a solid Cu_2SnS_3 thin film as,

$$Na_2S + H_2O \rightarrow 2Na^+ + HS^- + OH^- \tag{5}$$

$$HS^- + H_2O \rightarrow H_3O^+ + S^{2-} \tag{6}$$

$$2Cu^+ + Sn^{4+} + 3S^{2-} \rightarrow Cu_2SnS_3 \tag{7}$$

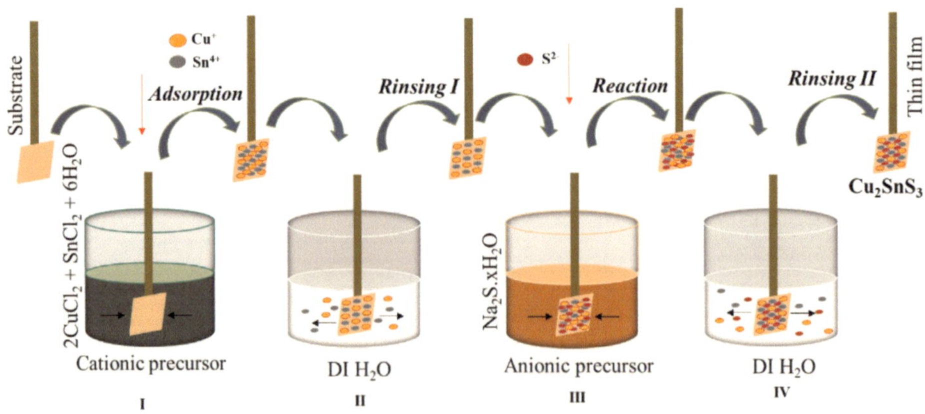

Figure 2.
Schematic representation of Cu_2SnS_3 thin film fabrication by SILAR technique [21].

In the last step of the process, the substrate was again dipped into the DI H_2O to remove the unwanted excessive particles to provide a uniform surface containing Cu_2SnS_3 thin film.

3. SILAR-facilitated material deposition: A summary

The deposition of the series of chalcogenide mainly metal oxides, sulfides, selenides, and tellurides films has been always on numerous attentions in the advancement of the SILAR since its launch. Currently, SILAR has become a broadly functional technique in the deposition of a huge variety of semiconductor thin films. For the simplicity of discussion, we have summarized most of the metals still synthesized as metal compounds by SILAR in **Table 1**.

A list of materials deposited using SILAR technique with their growth conditions with the required raw materials for the growth is summarized in **Table 2**. For the simplicity, the discussion is divided into four parts as of **Table 1**, as specified below:

3.1 Metal oxides

An increasing number of oxide materials deposited by SILAR have demonstrated high chemical, thermal, and expected stability that is one of the reasons to increase the popularity of oxide synthesis by SILAR. However, the technique of oxide synthesis is somehow difficult compared to sulfides, selenides, and tellurides due to the unavailability of the anionic precursors, which is the direct source of O^{2-} to form oxides. For example, in case of the synthesis of most of the binary metal oxides, H_2O, NaOH, and NH_4OH are used as anionic precursors with a mild thermal treatment of around (70 ~ 90) ^0C to activate the precipitation of hydroxides. On the other hand, the most common cationic precursors are mainly of metal thiosulfates, sulfates, chlorides, nitrates, etc., to provide metal ion adsorption on the substrate surface. Until today, Cu_xO, ZnO, TiO_2, and CdO are the most examined materials by SILAR. The investigation of Mn_3O_4, NiO, and Bi_2O_3 is also increasing [31–34]. Recently, both nanostructured Fe_2O_3 and Fe_3O_4 have been fabricated applying sulfate and chloride salts using NaOH as the anionic precursor *via* SILAR [30, 66]. But research on WO_3 [32], MgO [34], and SnO [67] fabrication is still rare. In case of ternary metal oxides, the SILAR deposition has been widely increased due to their ability to the additional modulate characteristics by controlling the composition of the materials. The synthesis of CST is discussed in the theory and mechanism section, which can be again done by two ways—a combined solution of both the deposited metal cations or, an alternating (one by one) fabrication of the two cations. A good technique to produce ternary metal oxides with excessive control on stoichiometry is to react one of the two metal ions by its own oxyanion. For example, $Bi(NO_3)_3$ and NH_4VO_3 react to fabricate $BiVO_4$ [37], as ammonium vanadates are extremely soluble in water, while the anticipated metal vanadates are not. Consequently, they precipitate out of solution as the

Oxides	Sulfides	Selenides	Tellurides
Mg, Ti, V, Mn, Fe, Ni, Cu, Zn, Mo, Ag, Cd, W, Bi, Sn	Cu, Zn, Cd, Ni, Sn, Pd, Mo, Ag, As, Bi, Mn, Sb, Fe, La	Cu, Sb, Cd, Bi	Cu, Cd, La

Table 1.
List of the metals still grown by SILAR technique.

Material	Precursors		pH (C, A)	Temperature, °C		Cycles	Rinsing	Dipping time, s (C, A)	Refs.
	Cationic	Anionic		Growth	Anneal				
				OXIDES					
Cu_2O	$Cu(S_2O_3)^-$	NaOH	—	70	—	10, 15, 20, 25, 30	No	5, 5	[24]
CuO	$CuSO_4$	NH_4OH	C:10	80	—	10, 20, 30, 40	H_2O	10, 15	[25]
ZnO	$ZnSO_4$	NH_4OH	—	100	200	100	No	2, 2	[26]
TiO_2	$TiCl_3$	NaOH	C: 3		400	100, 150	H_2O	20, 10	[27]
CdO	$CdCl_3$	NH_4OH	C:12	90	400	60, 70, 80, 90, 100, 110	H_2O	10, 20	[28]
Fe*x*O*y*	$FeCl_3$	NaOH	—	—	400	120	H_2O	10, 10	[29]
Mn_3O_4	Mn_2SO_4	NaOH	—	60		50	H_2O	10, 10	[30]
NiO	Ni $(NH_3)_4$	H_2O	—	90	200, 300, 400	100	H_2O	30, 7	[31]
WO_3	H_2O_4W	NaOH	—	80	400	25, 50, 75, 100, 125	H_2O	20, 10	[32]
Bi_2O_3	$Bi(NO_3)_2$+ $N(CH_2CH_3)_3$	H_2O	—	70	—	80	H_2O	10, 10	[33]
MgO	$Mg(NO_3)_2$	NH_4OH	A:11.5	90	400	—	No	30, 30	[34]
$Cu_2V_2O_7$	0.025 M $Cu(NO_3)_2$	0.025 M NH_4VO_3	—	75	—	5, 10, 20, 30, 60, 70, 80, 100	No	20, 20	[35]
Ag_3VO_4	$AgNO_3$	Na_3VO_4	—	70	100 ~ 350	5, 10, 15, 20	H_2O	20, 20	[36]
$BiVO_4$	$Bi(NO_3)_3$	NH_4VO_3	—	75	450	5, 10, 20, 30	H_2O	20, 20	[37]
$Fe_2V_4O_{13}$	$Fe(NO_3)_3$	NH_4VO_3	—	75	400, 500	20	H_2O	20, 20	[38]
$ZnFe_2O_4$	$ZnCl_2$, $FeCl_2$	Aq. NH_3	—	—	550	—	H_2O	30, 30	[39]
BiOI	$Bi(NO_3)_3$	KI	—	—	—	5, 15, 30, 45	H_2O	10, 10	[40]
$NiMoO_4$	$NiCl_2$ + NH_4OH	Na_2MoO_4	C:12.2	70	300	30	H_2O	20, 20	[41]
$FePO_4$	$FeCl_3$	NaH_2PO_4	—	—	—	25	H_2O	10, 10	[42]

Material	Precursors		pH (C, A)	Temperature, °C		Cycles	Rinsing	Dipping time, s (C, A)	Refs.
	Cationic	Anionic		Growth	Anneal				
				SULFIDES					
CuS	$CuCl_2$	Na_2S	C: 3, A:12	RT	200, 300, 400	40	H_2O	30, 30	[43]
Ag_2S/ Ag_2S_3	$AgNO_3$	$Na_2S.xH_2O$	—	55	—	12	CH_3OH	30, 45	[44]
Ag_2S	$AgNO_3$	$SC(NH_2)_2$	C:8, A:6	—	—	30	DD, H_2O	12, 10	[45]
ZnS	$ZnCl_2$	Na_2S	C:5, A:12	RT	200, 300, 400	40	H_2O	20, 20	[43]
CdS	$Cd(NO_3)_2$	Na_2S		—	—	9,12,15,18	H_2O	60, 60	[46]
NiS	$NiSO_4$	Na_2S	C:8, A:10	RT	—	150	H_2O	60, 60	[47]
SnS	$SnCl_2$	Na_2S	C:1.8, A:9.5	7	—	—	H_2O	20, 20	[48]
PbS	$Pb(NO_3)_2$	Na_2S	—	10, 30, 50	—	—	CH_3OH	60, 60	[49]
MoS_2	$(NH_4)_6Mo_7O_2$	Na_2S	C: 3, A:13.5	27	—	100	H_2O	25, 25	[50]
As_2S_3	As_2O_3 + EDTA	$Na_2S_2O_3$	C: 1, A: 6	80	—	10, 15, 20, 25, 30, 35, 40, 45, 50	DI H_2O	15, 15	[51]
Bi_2S_3	$Bi(NO_3)_3$	Na_2S	—	75	200	30	2-$CH_3OC_2H_5$ & CH_3OH	15, 15	[52]
MnS	$MnCl_2$	Na_2S	C:5.5, A:12	RT	—	60, 80, 100, 120, 140	H_2O	30, 30	[53]
Sb_2S_3	$SbCl_3$	$Na_2S_2O_3$	—	50	—	25, 30, 33	H_2O	20, 20	[54]
FeS	$FeSO_4$	Na_2S	—	27	—	40	H_2O		[55]
Fe_xS_x	$Fe(NO_3)_2$	Na_2S	—	27	200	20	H_2O	30, 30	[56]
CoS	$CoSO_4$	Na_2S	C:8–10 A: 12	27	—	25	H_2O	50, 50	[57]

Material	Precursors		pH (C, A)	Temperature, ^0C		Cycles	Rinsing	Dipping time, s (C, A)	Refs.
	Cationic	Anionic		Growth	Anneal				
La_2S_3	La_2O_3	$Na_2S_2O_3$	2 ± 0.1	RT	300	90	H_2O	30, 30	[58]
				SELENIDES					
Cu_3Se_2	$CuSO_4$	Na_2SeSO_3	C:2	RT	300	60	H_2O	20, 20	[59]
Sb_2Se_3	$SbCl_3$	Se^{2-}		RT	80	10	C_2H_5OH		[60]
CdSe	$CdCl_2$. H_2O	Na_2SeSO_3	C: 8, A:11.3	RT		30, 40, 50 60	H_2O	60, 60	[61]
Bi_2Se_3	$Bi(NO_3)_3$ + N $(CH_2CH_2OH)_3$	Na_2SeSO_3	C: 8, A: 8.5	27	—	150	H_2O	30, 30	[62]
Sb_2Se_3	$K(SbO)C_4H_4O_6$	Na_2SeSO_3	C:3, A:8.5	27	—	350	H_2O	40, 40	[62]
				TELLURIDES					
CdTe	$Cd(CH_3CO_2)_2$	Na_2TeO_3	C: 5, A: 11	27	—	910	H_2O	30, 30	[63]
Cu_2Te	$CuSO_4$	Na_2TeO_3	C:5, A: 9	27	—	60	H_2O	20, 20	[64]
La_2Te_3	$LaCl_3$+ CH_3COOH	Te pow. + HCl + HNO_3	C:2, A:1	RT	180	—	H_2O	20, 20	[65]

Table 2.
A list of materials deposited using SILAR technique with their growth conditions. (C, a: Cationic, anionic).

expected phase on the substrate surface. The other oxides such as Ag_3VO_4 [36], $BiVO_4$ [37, 68], $Cu_2V_2O_7$ [35], and $Fe_2V_4O_{13}$ [38] follow similar trends. Though bismuth oxyhalides, for instance, BiOI [69, 70] is not a pure oxide but have been synthesized by the SILAR using cationic precursor of bismuth nitrate and anionic precursor of KI [71].

3.2 Metal sulfides

The characteristic easiness of the procedure and wide-ranging obtainability of the anionic precursors afford metal sulfides the most fabricated materials by employing SILAR technique. The easiest way of sulfide thin film deposition is to use metal salts as cationic precursors and H_2O-soluble sodium sulfides or thiosulfates as anionic precursors. For example, NiS thin films could be fabricated using $NiSO_4$ (pH: 8) as cationic precursors and Na_2S (pH: 10) as anionic precursors even at room temperature [47]. Generally, the solubility product constants of used sulfide materials in water are higher than 10^{-20}, for instance, CuS: $\approx 10^{-36}$ and CdS: $\approx 10^{-27}$, which is the key force of the deposition of the expected metal sulfides. Among the sulfides, CuS [72], ZnS [43, 73], CdS [46], Ag_2S [44, 45], SnS [48], and PbS [49] have been broadly investigated mostly depending on the usage of chlorides and nitrates as the metal precursors and Na_2S as anionic precursors for the S^{2-} source. Moreover, the investigation on NiS [47], Bi_2S_3 [52] and MoS_2 [13, 50], As_2S_3 [13, 51], MnS [53] are growing fast, while in the case of CoS [57], La_2S_3 [58] studies are still infrequent. Moreover, core@shell-like SnS_2@Co_3S_4, ternary ($NiCo_2S_4$), and quaternary (Cu_2BaSnS_4) films as well as nanocomposites of CdS and Bi_2S_3 have been reported with their potential applications.

3.3 Metal selenides

In most of the cases, SILAR fabrication of the metal selenides has been directed through a solution of chloride, nitrate, or sulfate functioned as cationic precursor consisting of the anticipated metal, and a solution of Na_2SeSO_3 worked as the anionic precursor consisting of the source of Se^{2-} to form selenides. For example, 0.2 M $CdCl_2$. H_2O (pH: 8) reacts with 0.1 M Na_2SeSO_3 (pH: 11.3) to fabricate CdSe thin film [61]. Based on requirements, sometimes NaHSe, ethanolic $NaBH_4$, or Na_2Se can also be used as an anionic precursor to fabricate metal selenides. The SILAR deposition of Cu_3Se_2, Sb_2Se_3, Bi_2Se_3, and CdSe were studied and investigated mainly at room temperature avoiding thermal treatment during the sample growth, which was always maintained at (70–90) ^{0}C in the traditional cases.

3.4 Metal tellurides

The minimum studied materials among the chalcogen members *via* SILAR technique are metal tellurides because of the unavailability of the appropriate anionic precursors. Na_2TeO_3 or ethanolic Te or TeO_2 with $NaBH_4$ is the mostly used anionic precursor performed as the source of Te^{2-} to form tellurides. For example, 0.1 M $CuSO_4$. $5H_2O$ (pH: 5) and 0.05 M Na_2TeO_3 (pH: 9) react at an ambient temperature to synthesize Cu_2Te film. Till now, CdTe, Cu_2Te, La_2Te_3, Cu_7Te_4, and Bi_2Te_3 thin films were fabricated and investigated *via* SILAR technique having potential uses in case of radiation detectors, photovoltaics, and thermo-electric devices [63–65]. More scientific research is expected to understand and control the characteristics of such fabricated films to build outstanding optoelectronic devices.

4. Recent advances by SILAR

The optoelectronic properties of SILAR grown thin films have been demonstrated in many more applications, for example, supercapacitors, photovoltaics, photoelectrochemical water splitting, gas sensors, and many more. The technique seems to be simpler and represents an efficient way to fabricate devices. Three potential applications such as supercapacitors, photovoltaics, and photoelectrochemical water splitting will be discussed in the following section.

4.1 Supercapacitors

The rapid progress in state-of-the-art tools has guided to a profound reliance on energy storage devices. Satellites, electric vehicles, laptops, cellphones, and sensors need some species of energy storage to function properly. The lead-acid battery was the first device, discovered around the 1800s, and most common storage energy till today. Supercapacitors, another promising energy storage device, well known as electrochemical capacitor or ultracapacitor creates a gap bridging role between conventional capacitors and batteries [74]. They can offer $1 \sim 2$ orders of higher magnitude of power density than rechargeable batteries as well as supply much more energy than traditional dielectric capacitors.

A supercapacitor works following two-charge storage mechanisms: (i) surface ion adsorption such as electric double-layer capacitance (EDLC) and (ii) redox reactions such as pseudo capacitance. Supercapacitors reveal an extraordinary set of features in comparison with batteries, for instance, high-power density, low maintenance cost, reliable cycling life, fast rates of charge or discharge, and safe operation as well as offer versatile powering solutions to many appeals ranging from portable consumer electronic appliance and electric automobiles to large-scale smart utility grids. Nevertheless, carbon-based EDLC supercapacitors show very low energy densities, which are limited through the finite electrical charge separation at the interface of electrolyte and electrode materials, as well as the approachability of surface area. Consequently, efforts to surge the energy densities of supercapacitors have involved the application of better pseudo-capacitance electrode supplies, equipped by conducting polymers and nanostructured metal oxides, bearing the low cost of high-power density as well as chemical stability, which have the significance of phase changes and faradaic reactions in it [74].

Several types of metal oxides, sulfides, and tellurides have been used in supercapacitor device fabrication so far, by utilizing the ever-fast-growing technique SILAR as summarized in **Tables 3** and **4**. Initially the single metal oxides or sulfides such as CuO, NiO, $NiMoO_4$, WO_3, Bi_2O_3, Mn_3O_4, or MnS have been prepared by following SILAR technique and then tested for the supercapacitor behaviors to acquire the results of specific capacitance with their retention stability using cyclic voltammetry (CV) with the assistance of 3-electrode measurement system. Higher capacitance was attained at the lower scan rate and/or lower current density during such measurements and usually a relatively small quantity of electrochemical active material was developed atop of the working electrode. Moreover, the performance found using the 3-electrode system is higher than 2-electrode test cells, and the latter can be either a symmetric (S) or asymmetric (A) cell. Generally, in a symmetric cell both positive and negative electrodes are alike, whereas they are different active materials in an asymmetric cell.

Electrode	Scan voltage (V)	Electrolyte	Specific capacitance, Fg^{-1}	Current density	Capacitance retention	Ref
MnO_2	−0.1 to +0.9	1 M Na_2SO_4	314	5 mVs^{-1}	90% 10,000	[75]
Mn_3O_4	−0.2 to +1.2	1 M Na_2SO_4	786.2	5 mVs^{-1}	71% 1000	[76]
MnS	−0.15 to +0.55	6 M KOH	828	5 mVs^{-1}	85% 5000	[77]
NiO	−1.0 to +0.45	1 M KOH	1341	2 mVs^{-1}	90% 1000	[78]
CuO	−0.3 to +0.5	1 M Na_2SO_3	476	1 $mAcm^{-2}$	89% 1500	[79]
CuO	−0.8 to +0.4	1 M Na_2SO_4	554	1 Ag^{-1}	92% 4000	[25]
Bi_2O_3	−0.8 to +0.3	1 M Na_2SO_4	329.6	5 mVs^{-1}	72% 3000	[33]
WO_3	−0.7 to +0.4	1 M Na_2SO_4	266	10 mVs^{-1}	81% 1000	[32]
Fe_2O_3	−0.6 to +0.1	1 M NaOH	178	5 mVs^{-1}	— —	[24]
Fe_3O_4	−1.4 to −0.3	1 M KOH	488	1 $mAcm^{-1}$	— —	[29]
$NiCo_2O_4$	−0.1 to +0.5	3 M KOH	1936	5 mVs^{-1}	95% 1000	[80]
$NiMoO_4$	−0.1 to +0.4	2 M KOH	1180	1 Ag^{-1}	27% 1000	[81]
$NiMoO_4$	−0.6 to +0.6	2 M KOH	1853	1 Ag^{-1}	65% 2500	[41]
SnO_2-RuO_2	−0.2 to +0.6	0.5 M H_2SO_4	1010	1 $mAcm^{-2}$	85% 2500	[82]
SnS_2@Co_3S_4	0 to +4.5	—	1580	1 mVs^{-1}	95% 3000	[83]
NiO/MWCNTs	−0.1 to +0.5	2 M KOH	1727	5 $mAcm^{-2}$	91% 2000	[84]
Ni-Co-Ox/C-black	−0.2 to +0.4	1 M KOH	1811	0.5 $mAcm^{-2}$	92% 8000	[85]
Bi_2S_3: PbS/MWCNTs	−0.1 to +0.7	0.5 M Na_2SO_4	864.2	2 mVs^{-1}	68% 4000	[86]

Table 3.
Properties of various electrode materials deposited by SILAR for electrochemical capacitors by 3-electrode system.

Not only binary, but also ternary or even doped metal oxides, sulfides or tellurides were synthesized *via* SILAR for supercapacitor device application. For example, $ZnCo_2O_4$ and $ZnFe_2O_4$ were synthesized via SILAR technique from binary cationic solutions in the presence of Zn and Co (or Fe) precursors and demonstrate high energy density of 9.67 and 28 Wh kg^{-1} as well as power density of 1451 and 7970 W kg^{-1}, respectively. La_2S_3 and La_2Te_3 with mesoporous pine-leaf structure prepared with SILAR showed 35 and 60 $WhKg^{-1}$ energy density and power density of 1260 and 7220 WKg^{-1}, respectively. A flexible $La_2Te_3|LiClO_4\text{-}PVA|La_2Te_3$ supercapacitor cell was further fabricated and is represented as in **Figure 3**.

Hybrid supercapacitors, EDLC and pseudo-capacitance, build of charge storage mechanisms reduce the superior features of the device. On the other hand, among the other EDLC electrode materials multiwalled carbon nanotubes (MWCNTs) fascinated major interest due to their favorable features such as high surface area and mesoporous network, good mechanical strength and flexibility, excellent electrical conductivity, and chemical stability. The facile synthesis of composites of metal oxides with carbon materials was facilitated by SILAR as well. For instance, the fabrication of NiO/MWCNTs nanohybrid thin films *via* SILAR and the specific capacitance was as high as

Electrode Materials	S/A	Scan Voltage (V)	Electrolyte	Power density (WKg^{-1})	Energy density ($WhKg^{-1}$)	Capacitance retention		Ref
MnO_2	S	−0.1 to +0.9	1 M Na_2SO_4	1100	4.2	Stable	10,000	[30]
La_2S_3	S	−1.2 to 0	1 M Na_2SO_4	1260	35	78%	1000	[58]
La_2Te_3	S	−1.2 to 0	1 M $LiClO_4$/PC	7220	60	82%	1000	[65]
$ZnCo_2O_4$	S	0 to 1	PVA-KOH	1451	9.7	69%	2000	[87]
$ZnFe_2O_4$	S	0 to 1	PVA-$LiClO_4$	277	4.5	66%	1000	[88]
$ZnFe_2O_4$	A	0 to 1.6	6 M KOH	7970	28	74%	3000	[39]
Bi_2S_3: PbS/ MWCNTs	S	0.6 to 1.2	0.5 M Na_2SO_4	830	13.36	68%	4000	[86]
MnS@NF/rGO@NF	A	0 to 1.6	PVA–KOH	1280	34.1	87%	2000	[77]
Y-doped Sr.$(OH)_2$	S	0 to 1.5	PVA-Na_2SO_4	2270	3.7	61%	1000	[89]
In: $WO_3/Cu_7Te_4/$ Bi_2Te_3	S	0 to 1	Polysulfide	1700	18.85	—	—	[90]

Table 4.
Supercapacitors performance of SILAR grown films measured by 2-electrode system.

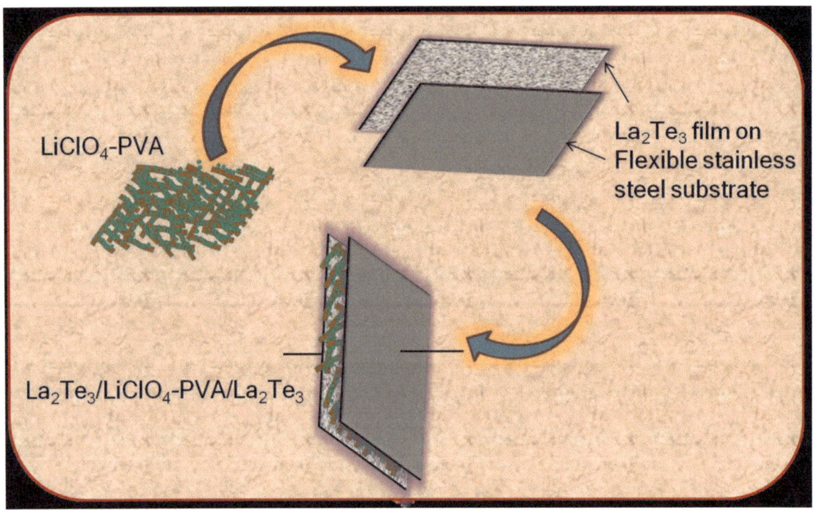

Figure 3.
Schematic diagrams of $La_2Te_3|LiClO_4$-$PVA|La_2Te_3$ supercapacitor device [65].

1727 Fg^{-1} and current density 5 $mAcm^{-2}$ with 91% retention ability after 2000 cycles as demonstrated in the **Figure 4** [84]. Moreover, an analogous synthesis style was employed to NiCoO*x*/Carbon-black hybrid thin films accomplishing coatings with a high specific capacitance of 1811 Fg^{-1} at 0.5 $mAcm^{-2}$ [85].

Therefore, SILAR is a unique as well as multipurpose technique to fabricate thin films for supercapacitor device application with superior power and energy densities in comparison with other available and more conventional deposition techniques, justifying the quality of the SILAR growth thin films.

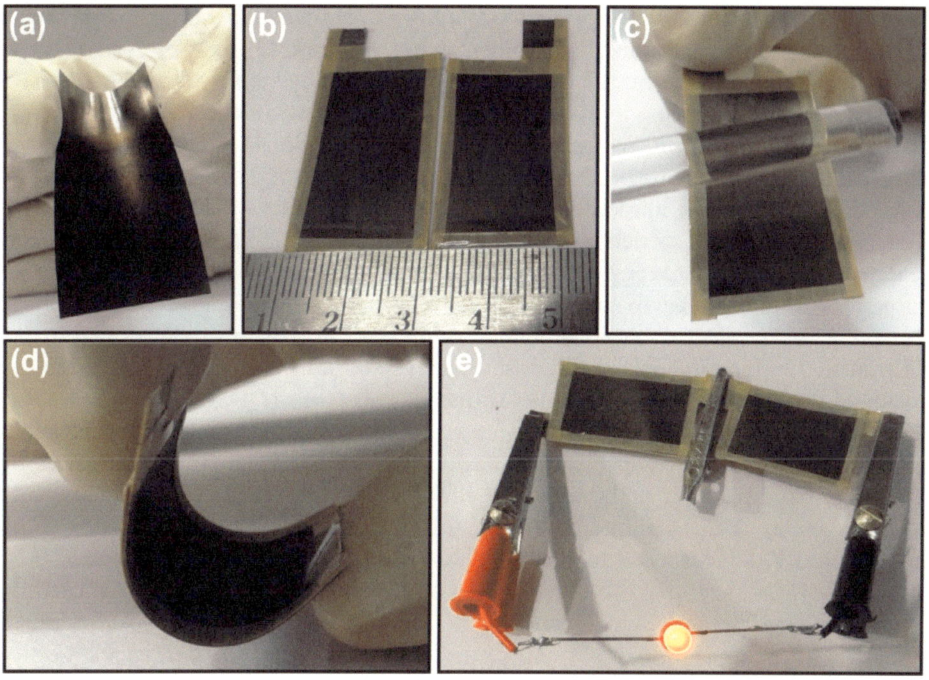

Figure 4.
Assembly of highly flexible symmetric NiO/MWCNTs-NiO/MWCNTs nanohybrid device: (a) image of flexible NiO/MWCNTs thin film deposited on the stainless steel, (b) NiO/MWCNTs nanohybrid thin film electrodes with closed ends (2 × 3 cm² area), (c) coating of the electrode by PVA/LiClO$_4$ gel electrolyte, (d) flexible supercapacitor built under the ~1 ton pressure through sandwiching the two-gel electrolyte coated electrodes, (e) two flexible supercapacitors in series can successfully light a LED [84].

4.2. Solar cells

Photovoltaic (PV) is a simple device, which promotes the direct conversion of light radiation into electrical energy by following the photovoltaic effect [91]. The discovery of such a device for the conversion of sunlight radiation directly into electricity was first carried out during the late 1800s. C. Fritts first demonstrated the solid-state PV by fabricating a thin layer of Au on Se semiconductor material [92]. At American Telephone and Telegraph Bell Laboratory the modern PV was discovered by Ohl in 1946 [93] but demonstrated by Chapin, Fuller and Pearson in 1954 [94]. The cell was fabricated by single-crystal Si wafer having an efficiency of ~5%.

At present for feasible use, extensive research is going for efficiency enhancement of solar cells, as the efficiency of solar cells is one of the very vibrant parameters to promote this technology. Over the years, the efficiency of single crystal-Si solar cells has shown a sound development. In 1950s, it was only 15% and nowadays it is improved to around 26.7% [95]. The commercial efficiency of Si solar cell is approaching in between 12% and 15%, while the theoretical Shockley-Queisser (SQ) limit energy conversion efficiency is of around 28% [96]. The PV cell and module market was mostly occupied on first-generation Si-based cells until 2004, for example, sc- and poly-Si cells, which covered about 85% of the overall international PV modules market. In the meantime, thin film cells or second-generation PV have exhibited great advantages, for instance, the ease of large area fabrication and usage of minimum materials, though their market share was much smaller in comparison with the Si-cells [97]. After 2005, the developments were spurred by the sharp increase in the country's implementation of solar energy due to the rapid advancement of the PV production industry in China. The price of PVs is generally supplemented by the strict requirement for fabricating high-purity materials such as GaAs and Si, or the rare-earth elements such as CIGS. The element, In, is rare and can be certainly exhausted, which might affect the prospect of such PVs. Later, CdTe thin-film PVs have increased longing in the market of South-Eastern countries. But Cd has a serious environmental distress, which is due to its high toxicity [98, 99]. For example, chronic Cd exposure breeds an extensive acute and chronic effects in humans [100, 101]. Moreover, Cd is a rare earth element and it will also generate a higher cost within the demand in future. Further, a third generation of devices has newly developed in addition to the thin-film solar cells, based on fresh organic materials such as dye-sensitized solar cells (DSSC) [102], quantum dot solar cells (QDSC), perovskites, bulk heterojunctions, having innovative device architectures with the usage of multiple exciton generation, upconverting layers, and others. Though, organic materials-built PVs have small life spans as the nature of the materials, for example, thermal stability [103] or concerns of electrolyte-based variability [104]. Inherently, a mostly striking new field of PV devices using metal oxide (MO) semiconductors has performed [105]. Atop of the MO thin films, favorable next-generation PV cells such as exciting thin absorber cells [106], DSSC [107], and QDSC [46, 108] are built as they are promising applicants for being stable, eco-friendly, and ultra-low-cost PV materials.

SILAR accounts itself directly into the third generation, by affording ultra-thin, compositionally fabricated by layers of several semiconductors that could be subjugated in a diversity of device architectures. Besides Si, most of the absorbers in PVs are conventionally II–VI, III–V semiconductors, or organic polymers, or small molecules, or perovskites. Still, the number of metal oxides is not adequate, which can be effectively used as absorber layers. Consequently, research into SILAR-grown light absorber layers for PV device applications has been aimed mainly on selective transition metal oxides, sulfides, and selenides. These materials have drawn incredible interest in technological and scientific research due to their unique optical, electrical, and mechanical properties in the past few decades [109–111]. Nevertheless, there are some examples of SILAR grown metal oxides, sulfides, and selenides applied in different types of solar cells such as thin film, DSSC, perovskite, and QDSC as summarized in **Table 5**. The layers were used not only to achieve high efficiency, but also served diverse roles inside the PVs such as light absorber, selective charge transport (electrons or holes), and passivation.

In many cases, the core light-absorbing layers, within the solar cells, fabricated *via* SILAR have been investigated. For example, in ITO/CdS/PbS/C heterojunction solar cell, n-type layer, CdS thin films were deposited by CBD on transparent conductive oxide (ITO) substrates, whereas PbS film by SILAR using different deposition cycles, 15, 20, 30, 40 and 60 to obtain different thicknesses, showed that 40 cycles PbS film has a greater photovoltaic conversion efficiency [115]. In another study, *p*-type CuO was utilized as photo-absorber in the p-CuO/n-Si heterojunction cell [123], where a vibrant role of the SILAR deposition was observed in the overall device performance, depending mainly on the concentration of the copper precursor solution. In an all-oxide solar cell, NiO/Cu_2O or CuO/ZnO/SnO_2, both Cu_2O and CuO fabricated were examined as light absorber fabricated by SILAR and the hole transporting layer (*p*-type NiO), buffer layer (ZnO) as well as *n*-type SnO_2 were deposited by sol-gel method [124]. The cell having Cu_2O showed better performance than CuO, which is due to the reduced conductivity, mobility, and carrier concentration of CuO. However, the study showed an overall

Absorber	Cell	Type	Current density	Circuit voltage	Fill factor	η %	Refs.
			J_{sc}, $mAcm^{-2}$	V_{oc}, mV	FF, %		
Cu_2O	Cu_2O/Pt: FTO	Thin film	3.29	0.35	30	0.35	[24]
Cu_2O	$Cu_2O/MAPbI_3$/PCBM	Perovskite	16.52	0.89	56	8.23	[112]
ZnO	ZnO/Pt: FTO	DSSC	4.04	0.44	33	0.59	[102]
ZnO	ZnO/Pt: FTO	DSSC	5.43	0.42	32	0.70	[113]
TiO_2	FTO/TiO_2/N3/I^- - I_3^-/Pt: FTO	DSSC	8.17	0.42	34	1.16	[27]
TiO_2	FTO/TiO_2/spiro OMeTAD/Au	Perovskite	18.05	0.81	66.3	~10	[114]
CuS, Ag_2S	CuS/Sn: Ag_2S	Thin film	8.3	0.66	52	2.85	[44]
CdS, PbS	ITO/CdS/PbS/C	Thin film	0.87	0.14	31.4	0.13	[19]
Bi_2Se_3	n-Bi_2Se_3/0.1 M polysulfide/C	Thin film	0.52	0.12	43	0.03	[115]
Sb_2Se_3	FTO/TiO_2/Sb_2Se_3/electrolyte/Pt	SSC	4.19	0.57	51.66	1.22	[60]
CdS	FTO/TiO_2/CdS QD/I^--I_3^-/Pt: FTO	QDSC	3.81	0.76	61.9	1.80	[46]
SnS	Gr/TiO_2/SnS QD/I^--I_3^-/Pt: FTO	QDSC	1.35	0.52	42	0.30	[108]
CdSe	ZnO/CdS/CdSe QD/MnS/electro. / Cu_2S	QDSC	13.74	0.60	44	3.70	[116]
CdSe	FTO/TiO_2/CdSe/ Na_2S: S/CoS_2	QDSC	16.86	0.54	39	3.53	[117]
CdS: Mn	Mn-d-CdS/CdSe/S^{2-}-S/RGO/Cu_2S	QDSC	20.7	0.56	47	5.42	[118]
CdSe	FTO/TiO_2/CdS/CdSe/S^{2-}-S_n^{2-}/Cu_2S	QDSC	13.64	0.56	51	3.89	[119]
PbS: Hg	FTO/TiO_2/PbS: Hg QD/S^{2-}-S_n^{2-}/ Cu_2S/Brass	QDSC	30	0.40	46.8	5.6	[120]
PbS	FTO/TiO_2/PbS QD/ S^{2-}-S_n^{2-}: CH_3OH/ Cu_2S	QDSC	18.34	0.43	50.86	4.01	[121]
SnSe	FTO/TiO_2/SnSe QD/S^{2-}-S_n^{2-}/Cu_2S / Brass	QDSC	6.94	0.57	19	0.78	[122]

N.B. *N3: ethanolic 0.3 mM cis-Bis (isothiocyanato) bis (2,2′-bipyridyl-4,4′-dicarboxylato ruthenium (II).*

Table 5.
SILAR growth PV cells demonstrating with the cell properties.

efficiency over 1%. In a different study, heterojunction solar cells have been fabricated between layers of p-type CuS and n-type Ag_2S deposited *via* SILAR method and Sn^{2+} and Al^{3+} heterovalent dopants are introduced in Ag_2S so that Fermi energy of the semiconductor can be modified to alter the band diagram of pn junctions. The Sn^{2+}-doped Ag_2S resulted in better solar cell parameters with an efficiency of 2.85% as compared to that based on Al^{3+}-doped Ag_2S, which consists of many defect states due to mismatch in ionic radii of the cation and the dopant ions [44].

Further, metal oxides were worked in charge transportation in between different layers in solar cells, and both electron transport layer (ETL) and hole transport layer (HTL) can enhance the performance of PVs. Since the early 1990s, TiO_2 is one of the key materials used as ETL owing to its wide popularity in DSSC [120]. In a recent study, TiO_2 nanocrystalline film was directly deposited using SILAR at 90°C for perovskite solar cell applications and used as an ETL [114]. Due to the fast charge transport, kinetics and slow charge recombination process of the TiO_2 ETL synthesized from the solutions of $TiCl_4$ and hot $K_2S_2O_8$, with subsequent annealing at 450°C, advances the efficiency of the cell to around 10%. Further, a couple of studies showed the deposition of TiO_2 layers from solutions of $TiCl_3$ and NaOH [27, 125] followed by annealing at 400°C, as ETLs in DSSC with the modest efficiency of just over 1%. Other SILAR-fabricated layers used as ETLs in PVs consist of ZnO and ZrO_2 as interfacial layer attached to porous TiO_2, both demonstrated performance in DSSC [102, 126].

In this study, Cu_2O thin films were introduced as a HTL in a planar perovskite solar cell and successfully enhanced the efficiency of the cell to around 8.23%, as shown in **Figure 5**(a-c). The Cu_2O films were deposited *via* SILAR by followed the complexation reaction of copper and ammonia with H_2O_2 [112]. The methylammonium lead triiodide ($MAPbI_3$) perovskite layer is sandwiched between a *p*-type Cu_2O HTL layer and another *n*-type PCBM (phenyl-C61-butyric acid methyl ester) ETL layer, respectively. The Cu_2O films demonstrated suitable band structure after annealing at 170°C and boosted device performances better than conventional sol-gel-deposited NiO and Cu-doped NiO hole transport layers, confirming the quality of the SILAR-Cu_2O.

In a report of 2009, Lee et al. showed a novel technique for preparing selenide (Se^{2-}) by the SILAR process in pursuit of efficient QD-sensitized solar cells atop of mesoporous TiO_2 photoanodes. After several optimization of the QD-sensitized TiO_2 films *via* regenerative photoelectrochemical cells in presence of a cobalt redox couple [$Co(o\text{-}phen)_3{}^{2+/3+}$], with a final layer of CdTe, the overall efficiencies of the was recorded around 4.2% at 100 Wm^{-2} [127]. To find the answer to a question, "How does a SILAR CdSe film grow?" Becker et al. tuned the deposition steps to suppress interfacial charge recombination in FTO/TiO_2/CdSe/Na_2S: S/CoS_2 cell by showing an efficiency of 3.53% [117]. Recently, in another report, SILAR- and CBD-grown CdSe-sensitized TiO_2 solar cells were examined concentrating on the influences of two commonly used QD deposition techniques [119], and atop of pre-accumulated CdS seed layers, a successful CdSe deposition was performed. The PEC of both the cells has been recorded as 4.85%, for CBD grown CdS/CdSe cell, whereas for SILAR grown cell the value was 3.89%. One research group enhanced the PCE of CdS/CdSe/S^{2-}-S/RGO/Cu_2S cell to 5.4% by employing Mn^{2+} doping of CdS *via* SILAR method [118], whereas another group reported on a PbS: Hg QD-sensitized solar cell by Hg^{2+} doping into PbS employing similar deposition technique and showed an unprecedentedly high J_{SC} of 30 mA/cm^2 with the PEC of 5.6% [120]. More studies are ongoing with great efforts to find new alternative, clean, and environment-friendly energy resources due to the increasing demands.

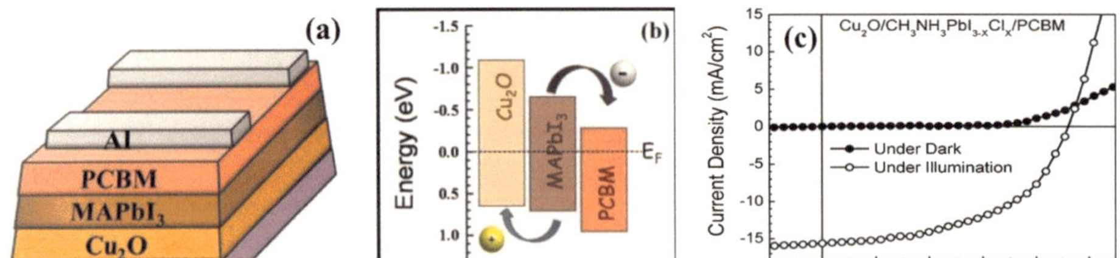

Figure 5.
(a) Cell structure, (b) schematic energy level diagram; the dashed line represents the Fermi energy after contact (c) current-voltage characteristics of under dark and a white light illumination condition of $Cu_2O/MAPbI_3/PCBM$ heterojunction [112].

4.3. Photoelectrochemical water splitting

Hydrogen energy is a key issue to cope with the present global energy crisis and environmental complication exploiting clean and inexhaustible energy [128]. Photoelectrochemical (PEC) water splitting is a promising technique to create hydrogen fuel by utilizing solar energy. Within the nonstop efforts in developing efficient photoelectrodes, the major challenge researchers presently face is to explore cost effective, nontoxic, and earth-abundant photoelectrodes with high efficiency [129]. In a recent study, Ag/Ag_2WO_4 was fabricated on ZnO nanorods using 0.05 M $AgNO_3$ and 0.05 M Na_2WO_4 as the cationic and anionic precursors, respectively, by following SILAR technique and the composite material demonstrated outstanding performances in PEC water splitting with 3 $mAcm^{-2}$ at 1.23 V versus RHE in the presence of 0.1 M Na_2SO_4 electrolyte. Based on these results, a brief possible updated mechanism of the PEC activity was demonstrated by Adam et al. for the better understanding of the technique with **Figure 6** [130]. The development of PEC activity of the semiconductors was principally attributed to electrons and hole transfer at the interfaces of the photoelectrodes. The band edge potentials of the Ag/Ag_2WO_4 and the ZnO materials showed a significant role in the efficiency of growth and separation technique of the electron (e^-) and hole (h^+) pairs. The energy of valence band (E_{VB}) of ZnO and Ag_2WO_4 is calculated as +2.86 and + 3.03 eV, whereas energy of conduction band (E_{CB}) of them is projected as −0.34 and − 0.07 eV, respectively [131].

In sunlight, both the semiconductors absorb light and the electrons in the VB become excited up to a higher potential of −0.34 and − 0.07 eV for the ZnO and Ag_2WO_4, respectively. Consequently, due to high photon energy, within the semiconductor, the effective charge transfer process proceeds. Ag^0 nanoparticles (NPs) cause active separation of h^+ or e^- pairs upon the absorption of light owing to the surface plasmon resonance (SPR) effect. Electrons from the Ag NPs are transported to the CB of the Ag_2WO_4 and the ZnO, while holes persist in the Ag NPs. In the meantime, to occupy the vacant holes created by the plasmonic absorption, the

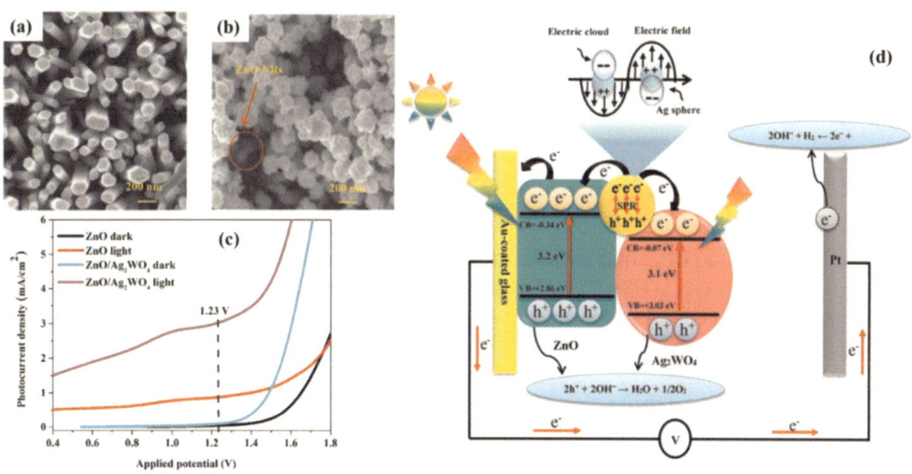

Figure 6.
(a) and (b) FM-SEM images of the ZnO NRs and the $ZnO/Ag/Ag_2WO_4$ heterostructure. (c) Curves of the ZnO NRs, and the $ZnO/Ag/Ag_2WO_4$ photo-electrodes under light and dark conditions using linear sweep voltammetry. Schematic diagram presenting the energy band structure and probable electron-hole separation as well as transportation in $ZnO/Ag/Ag_2WO_4$ heterostructure with the SPR effect [130].

photogenerated electrons in the CB of ZnO will be transported to the Ag NPs [132]. The photogenerated charge carriers can be proficiently separated to enhance the PEC performance by following this mechanism. Further, the photogenerated electrons will eventually reach at the Pt electrode (counter) and contribute to H_2 generation. Also, the photogenerated holes in the VB of Ag_2WO_4 and ZnO will contribute on O_2 production *via* H_2O oxidation. Hence, these outcomes validate the modification *via* Ag/Ag_2WO_4, which is an active technique to attain a high PEC activity by means of ZnO NRs arrays.

Like the above example, many studies based on SILAR had been devoted toward exploring the potential of semiconductor thin films as photoelectrodes for water splitting as shown in **Table 6** with their potential applications. In terms of low-cost, simplicity, and theoretically high solar to H_2 efficiency, PEC water splitting is much more favorable than solar photobiological, photochemical, and thermochemical generation of hydrogen [146]. The most investigated semiconductor materials include $BiVO_4$, Fe_2O_3, $CuCoO_2$, WO_3, and TiO_2 [147]. Other semiconductor materials such as Cu_2O [24, 148, 149], ZnO [150, 151], TiO_2 [152, 153], and CdO [28] were also produced using SILAR method but the PEC performances are quite low under visible light due to their wide bandgap.

5. Factors affecting SILAR deposition

A lot of research work has been done on the deposition and optimization of the SILAR thin films for optoelectronic device applications. Solution concentration, composition of precursors, the number of SILAR cycles, pH, annealing, and doping will absolutely affect the quality and quantity of thin films, which directly influence the cell performance. The effect of different parameters used in SILAR deposition on the performance of thin films is reviewed based on the contemporary research work.

5.1 Solution concentration

Solution concentration of the used precursor is one of the key factors in governing the properties as well as the performances of SILAR grown thin films. From a general viewpoint, depositing through a more concentrated solution results with bigger grain size and higher surface roughness during deposition. Consequently, thinner, smoother, and probably pinhole-free deposition can be attained using multiple SILAR cycles with a lower concentrated precursor solution.

With the increase of molar concentration (0.03, 0.05, and 0.1 M) of the cationic solution prepared by $Cd(CH_3COO)_2$ and H_2O_2, the surface morphology of the SILAR-deposited nanostructured CdO thin films was improved toward the crack free and homogeneous nature [154]. On the other hand, the structural change such as nanorods, nanoflowers, and nanoflakes morphologies was observed by altering only the concentration of anionic precursors NaOH (high, 0.05 M; moderate, 0.01 M; and low, 0.001 M) with fixed Zn precursor concentration (0.005 M) [155]. A comparative study of CdS films deposited by SILAR and CBD techniques revealed that the S/Cd ratio in the sample increases (0.83 to 1.04) for SILAR deposited films with the molar concentration of sulfur (1:1, 3:1, 5:1 and 7:1) in the starting solution increases, while it was almost constant ($\sim$0.80) for CBD films [156]. During the investigation of the effect of the molar concentration of pyrrole monomer on the electrochemical behavior

Films	Electrolyte	pH	Precursors			Rinsing	Current density	Light source	Refs.
			Cationic 1	Cationic 2	Anionic		$mAcm^{-2}$	lamp	
Ag_3VO_4	Na_2CO_3	11.4	Ag_3VO_4	—	Na_3VO_4	DI H_2O	0.07	Xe arc	[36]
$BiVO_4$	Na_2SO_3	7	$Bi(NO_3)_3$	—	NH_4VO_3	DI H_2O, H^+	~1.2	Xe	[37]
$BiVO_4$	0.1 M KPi	7	$Bi(NO_3)_3$	—	NH_4VO_3	HNO_3	1.5	Thorlabs	[68]
$Fe_2V_4O_{13}$	Na_2SO_3	9.2	$Fe(NO_3)_3$	$(NH_4)_6Mo_7O_{24}$	NH_4VO_3	—	0.028	300 $mWcm^{-2}$	[38]
TiO_2 NR/BiOI	Na_2SO_4		$Bi(NO_3)_3$	—	KI	—	0·40	Xe	[133]
SILAR-$BiVO_4$	Co-Pi	7	$Bi(NO_3)_3$	HNO_3	NH_4VO_3	—	2.3	Plasma	[134]
$Bi_2Mo_2O_9$@Bi_2MoO_6	K_3PO_4	7	$Bi(NO_3)_3$	$(NH_4)_6Mo_7O_{24}$	—	DI H_2O	0.16	Xe arc	[135]
$WO_3/BiVO_4$	H_2SO_4	—	$BiNO_3$	—	NH_4VO_3	DI H_2O	1.9	Xe	[136]
Fe_2O_3/TiO_2/FTO	NaOH	13.6	$FeCl_3$	—	NaOH	DI H_2O	0.4	Xe arc	[137]
Ti: α-Fe_2O_3	NaOH	13.6	FeCl3	$Ti\{OCH(CH_3)_2\}_4$	NaOH	DI H_2O	0.85	Xe arc	[138]
$CuCoO_2/BiVO_4$	Na_2SO_3	6.86	$Bi(NO_3)_3$	—	KI	—	3.3	Xe lamp	[139]
$ZnFe_2O_4/BiVO_4$	Na_2SO_4	7	$Zn(NH_3)_4^{+2}$	$FeSO_4$	NaOH	H_2O_2	1.46	Sim. Sun.	[140]
$CuWO_4$/BiOI	K_3PO_4	7	$Bi(NO_3)_3$	—	KI	$(CH_2OH)_2$	2.10	AM1.5 G	[141]
Ag/Ag_2WO_4/ZnO	Na_2SO_4	—	$Ag(NO)_3$	—	Na_2WO_4	DI H_2O	3	Xe	[130]
$Fe_2O_3/BiVO_4$	KH_2PO_4	7	$Fe(NO_3)_3$	—	NaOH	—	1.63	Xe	[142]
$TiO_2/BiVO_4$/NiFe-LDH	Na_2SO_4	—	$Bi(NO_3)_3$	—	NH_4VO_3	DI H_2O	0.018	Xe	[143]
$BiVO_4$/ZnO	Na_2SO_5	—	$Bi(NO_3)_3$	—	NH_4VO_3	—	1.72	Xe	[144]
$AgVO_3$/Mo: $BiVO_4$	Na_2SO_4	6.4	$AgNO_3$	—	NH_4VO_3	DI H_2O	1.93	Xe arc	[145]

Table 6.
SILAR growth films demonstrating the properties of photoelectrochemical water splitting at 1.23 V vs. RHE.

of highly pristine poly-pyrrole flexible electrodes, it was shown that among 0.025 M, 0.05 M, and 0.1 M pyrrole, the 0.1 M pyrrole exhibited excellent performance with specific capacitance as high as 899.14 Fg^{-1} at 5 mVs^{-1} in 0.2 M Na_2SO_4 showing retention stability of 61.5% even after 2000 cycles [107]. The SILAR synthesis, in such case, was performed on the stainless steel strips, which were firstly immersed in pyrrole precursor, followed by 30% H_2O_2 for 10 s each [107]. Further, studies on the use of the optimum precursor concentration of different Mn dopant (0.04 M, 0.075 M and 0.1 M) in CdS QDSSCs reveal that 0.075 M Mn-doped CdS can strongly enhance the incident photon to charge carrier efficiency (IPCE), due to the improved light harvesting, electron injection as well as charge collection efficiencies. As a result, the PCE of SILAR-grown Mn/CdS QDSC is up to 3.29%, which is much higher than that of QDSC without doping (2.01%) as well as other used concentration of Mn dopant under standard simulated AM 1.5 G, 100 mW cm^{-2} [157].

5.2 Effect of precursors

ZnO is one of the most investigated materials performed by SILAR technique. During fabrication, among the other properties precursor selection is one of the key requirements. In this study, the role of the precursor materials such as $Zn(CH_3COO)_2$, $ZnSO_4$, and $ZnCl_2$ on the properties of SILAR-deposited ZnO films were examined and the outcomes showed that the films fabricated by utilizing $Zn(CH_3COO)_2$ and $ZnSO_4$ precursors exhibited better optical properties than $ZnCl_2$. Besides, the crystallite sizes of all the fabricated samples were increased upon annealing [158]. On the other hand, the effect of four different precursors of $Zn(NO_3)_2$, $Zn(CH_3COO)_2$, $ZnSO_4$, and $ZnCl_2$ on structural, morphological, electrical and optical properties of AZO thin films using SILAR method was examined. After varying the different precursors, the significant effects on film crystallization, surface morphology, optical nature, and electrical resistivity of the deposited films were studied, in which chloride precursor demonstrated the best performance [159].

Sfaelou and co-workers studied the effect of the nature of three cadmium precursors such as $Cd(NO_3)_2$, $CdSO_4$, and $Cd(Ac)_2$ on the effectiveness of CdS SILAR deposition and measures the performance of sensitized solar cells and photo fuels. The CdS reflection spectra, load, and the size of CdS nanoparticles varied a lot from one precursor to the other as shown in **Figure** 7(a-e). The highest load and the largest nanoparticles were obtained in the case of $Cd(Ac)_2$, and the smallest in the case of $Cd(NO_3)_2$. And acetate-derived photoanodes provide more effective outcomes in the case of QDSSCs, while nitrate-derived precursors were more effective in the case of photo fuel cells as in **Figure** 7(f, g) [11]. In a similar but detailed study, Zhou and co-worker showed almost similar results showing a better performance of $Cd(Ac)_2$ over $Cd(NO_3)_2$ during the study of another CdS QDSCs as shown in **Figure** 7(h) [119]. Another recent investigation on the effect of different precursors such as $Mn(CH_3CO_2)_2$, $MnCl_2$, and $MnSO_4$ on electrochemical properties of Mn_3O_4 thin films prepared by SILAR method using 1 M Na_2SO_4 aqueous electrolyte exhibited the specific capacitance of 222, 375, and 248 Fg^{-1}, respectively, at 5 mVs^{-1} scan rate. Hence, the $MnCl_2$-derived Mn_3O_4 electrode showed a good electrochemical with maximum energy density of 17 $Whkg^{-1}$ and power density of 999 Wkg^{-1} at 0.5 $mAcm^{-2}$ current density showing retention stability of 94% after 4500 CV cycles [160]. Besides, a study of SILAR-deposited SnO_2 films showed improvement of the crystallite with solution molarity performed by using different precursor concentration of both cations and anions [161].

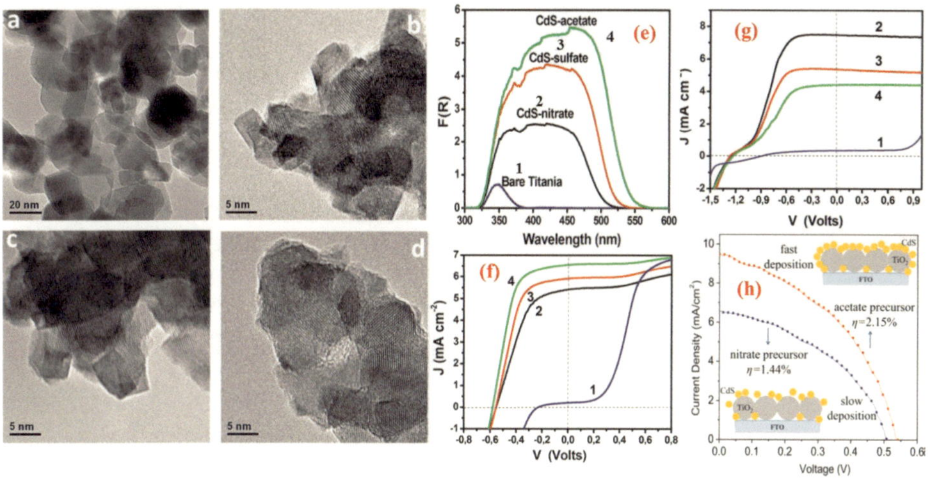

Figure 7.
TEM and HRTEM photographs of (a) pure titania and titania loaded with CdS deposited by using the three precursors: (b) $Cd(NO_3)_2$, (c) $CdSO_4$, and (d) $Cd(Ac)_2$. (e) Reflection spectra using $Cd(NO_3)_2$, $CdSO_4$, $Cd(Ac)_2$, and titania film without CdS. $J - V$ curves recorded with a (f) QDSSC and (g) photo fuel cell employing 1 cm^2 active photoanode and 2.25 cm^2 active cathode electrode [11]. (h) $J - V$ characteristics of CdS QDSCs fabricated by using acetate and nitrate precursors and measured under the illumination of one sun (AM 1.5, 100 mW/cm^2) [119].

5.3 Number of deposition cycles

SILAR technique involves the successive immersion of the substrate in anionic and cationic precursors following the substrate rinsing procedures in between. The deposition rate and the thickness of the required films can be simply controlled over a wide range by varying the deposition cycle and there are no boundaries on the substrate material, dimensions, or surface profile to be used, which in turn influence the properties such as crystallite size, surface morphology, and possibly light absorption. Nevertheless, overloading may consequence in delamination and fragmentation of the films owing to undesirable mechanical stress. The number of cycles optimization is therefore requisite to all SILAR system for the anticipated utilization.

Recently, lily flower-like ZnO structures were demonstrated by a group of researchers deposited by SILAR method [162]. In the study, lily flower-like morphologies were obtained when the deposition cycle number increases from 1 to 10 as shown in **Figure 8**(a-d). Another group, while studying the growth of porous $Fe_2V_4O_{13}$ films for photoelectrochemical water oxidation, the deposition cycle had directly altered the current density as shown in **Figure 8**(e). The highest photocurrent was achieved at the potential of 1.23 V *vs.* RHE for a $Fe_2V_4O_{13}$ film attained through 20 deposition cycles, which was chosen to improve the performance of the material further [38]. Further, Das and coauthors studied the influence of dipping cycle on SILAR-synthesized NiO thin film and observed that 40 cycle dipping NiO electrode provides highest specific energy of 64.38 $WhKg^{-1}$ with the highest specific power 2305 WKg^{-1}, by retaining fast electron transfer as well as admission of electrolyte ions much easily due to porous nanostructure of fabricated electrode [78]. Moreover, other efforts including the effect of immersion cycles on structural, morphology, and optoelectronic properties such as Ag_2S [163], CdO [164], ZnS [165] thin films were studied extensively, which make them desirable for optical coating as well as other optoelectronic applications.

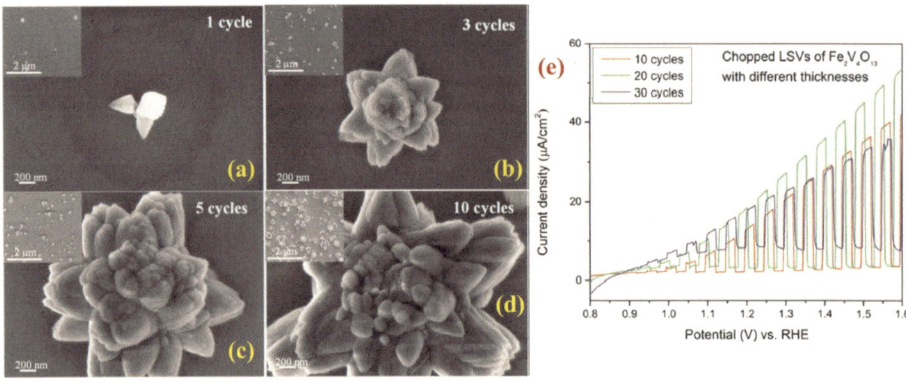

Figure 8.
(a-d) FE-SEM photographs of ZnO lily flower-like structures deposited by varying number of deposition cycles via SILAR method [162]. (e) Chopped LSVs of $Fe_2V_4O_{13}$ films with different number of cycles annealed at 500°C for 1 h in a buffer solution of pH 9.2 [38].

5.4 Impact of pH

By altering the pH of both cationic and anionic precursors, it is possible to tune the bandgap of thin films over a wide range for optoelectronic device applications. Preetha and co-workers investigated the effect of cationic precursor pH on optical as well as transport properties of SILAR-fabricated nanocrystalline PbS thin films. They successfully showed that the pH of the cationic precursor and in turn the size of the crystallites affect the optical and electrical properties of PbS thin films [166]. Besides, Sakthivelu and coauthors demonstrated a similar effect on ZnO thin films that the grain size of ZnO increased with the increase in pH of the precursor solution as represented in the SEM micrograms in **Figure 9**(a-e). The film deposited at pH =8.5 shows aggregated and non-uniform grains, while flower-like appearance appeared at pH =9. Later, at pH =9.5, bigger grains with hexagonal nanorods structure appeared and finally, at pH =10, the size of the nanorods increased further with the well

Figure 9.
SEM micrograms of ZnO thin films prepared at (a) pH = 8.5, (b) 9.0, (c) 9.5, (d) 10.0, and (e) 10.5 [167].

elongated nanorods sticking with each other. They also showed a decrease of bandgap from 3.29 to 3.09 eV with the increase of pH [167].

Moreover, both acidic and basic mediums can be used based on the requirement to deposit films as listed on **Table 7**. Farhad and co-author worked on Cu_2O thin film under pH range from ~2 to 8 [169, 170], while Gençyılmaz [171] and Visalakshi et al. [172] on CuO thin film under pH range from ~9 to 12 by showing promising electrical and optical properties. In another research, CdO was deposited using $Cd(CH_3COO)_2$ as a cationic precursor and thickness as well as bandgap tuning was effectively observed in the pH range of 11.3 to 12.5, in addition to NH_3 solution [173]. However, not only a wide range of study to understand the influence of pH on the thin films fabricated by SILAR, but also more device fabrications were performed under diverse pH conditions with excellent performance, as described in the application section.

5.5 Annealing

SILAR is currently in demand to maintain the high quality of films with high growth rate. Despite extensive efforts, the adsorption of complex agents from precursors, drop of rinsing time, and slow production rate for almost all kinds of films has still been a main disadvantage, which restricts its usage in semiconductor industry. New deposited films may have defects, for instance, oxide vacancies and hydroxide phase since the presence of hydroxide phase is unavoidable owing to an aqueous alkaline medium for fabrication [174]. However, annealing of films can minimize such defects and eliminates the hydroxide phase along with the recrystallization. As high-temperature annealing mostly induces a rise in crystallite size, and possibly alters in morphology, which is steady with thermally induced grain growth. Still, the bandgap does not allow a steady trend with annealing due to numerous factors coming into ground outlining the absorption onset: size of particles, existence of defects, stoichiometry as well as existence of oxygen vacancies, etc. [175].

Putri and co-authors studied annealing temperature effect on the photovoltaic performance of BiOI-based materials and showed that at 300°C temperature, the role of the device which consisted of $Bi_7O_9I_3$ attained three times higher efficiency than the annealed parent BiOI at 100°C. Hence, the structural tuning due to the addition of oxygen *via* annealing to BiOI structure had an influence on the photoelectrochemical cell [69]. Besides, Ashith and co-worker studied the effect of post-deposition annealing on the properties of ZnO films and demonstrated that the crystallite size of the films increased significantly after annealing. The annealed films further showed very high absorption in the UV region with marginal modification in bandgap. Both the crystallite size and optical absorbance were observed to rise proportionately with the annealing temperature [176]. In a separate study of annealing and light effect on structural, optical and electrical properties of CuS, and $Cu_{0.6}Zn_{0.4}S$ thin films grown by the SILAR demonstrated that the current increase with increasing light intensity and increasing rate in illuminated 500 Wcm^{-2} films were greater than the others that have annealed at 400°C [43]. Further studies are reported showing grain size increase after annealing or bandgap tuning are listed in the **Table 8**, including Cu_2O, ZnO, CuO, CdO, MgO, NiO etc.

5.6 Impact of doping

In order to have a maximum number of carriers to take part in the functioning, a material with low activation energy is necessary so that electrons can easily jump from

Film	pH	Precursors		To Adjust pH	Thickness (nm)	Crystallite size (nm)	Dislocation density ($\delta \times 10^{15}$ lines m^{-2})	Strain ($\varepsilon \times 10^{-3}$lines^{-2} m^{-4})	Texture coefficient TC (hkl)	Bandgap (eV)	Resistivity $\times 10^3 (\Omega cm)$	Refs.
		Cationic	Anionic									
Cu_2O	2.35	[Cu $(S_2O_3)]^-$	*2 M NaOH	SA	340	17	3.460	2.28	0.744	2.05	0.21	[169]
	3.45		*2 M NaOH	AA	729	21	2.268	2.11	0.756	2.10	0.18	
	4.50		2 M NaOH	AA	800	15–22	—	—	—	2.30	72	[170]
	5.10		2 M NaOH	AA	1000		—	—	—	2.28	103	
	6.20		2 M NaOH	AA	1800		—	—	—	2.43	742	
	7.33		*2 M NaOH	~	1130	18	3.086	2.20	0.768	2.15	0.37	[169]
	7.33		#1 M NaOH	~	336	13	5.917	2.90	0.561	2.16	0.18	
	7.95		2 M NaOH	~	1477	15–22	—	—	—	2.42	21.9	[170]
CuO	9.0	[Cu $(NH_3)_4]^{2+}$	H_2O	~	42	37.4	0.714	9.26	1.21	1.61	—	[171]
	9.5		H_2O	~	67	22.4	1.992	15.4	1.17	1.49	—	
	10.0		H_2O	~	85	22.9	1.906	15.1	1.07	1.49	—	
	10.0		H_2O	SA	520	14	5.102	2.31	1.05	2.17	6.5	[172]
	10.5		H_2O	SA	590	21	2.268	1.63	1.34	2.07	5.5	
	11.0		H_2O	SA	680	27	1.372	1.48	1.65	2.02	4.0	
	11.5		H_2O	SA	770	30	1.111	1.33	1.71	1.99	4.25	
	12.0		H_2O	SA	820	36	0.716	1.24	1.83	1.89	4.5	
CdO	11.3	Cd $(CH_3COO)_2$	H_2O	NH_3	520	14.59	5.44	2.48	2.52	3.26	—	[173]
	11.7		H_2O	NH_3	1390	16.86	4.09	2.15	1.89	3.21	—	
	12.1		H_2O	NH_3	1820	17.22	3.91	2.10	1.90	2.67	—	
	12.5		H_2O	NH_3	2360	17.38	3.84	2.08	1.71	2.35	—	

**OP = Optimized precursor (2 M NaOH), NOP = Non optimized precursor, SA = Sulfuric acid and AA = Acetic acid.*

Table 7.
Properties of thin films deposited by varying solution pH via *SILAR method [168].*

Temperature, ^oC		Precursors		Cycle	Time		Crystallite	Bandgap	Ref
Annealing	Growth	Cationic	Anionic		Dipping (sec)	Annealing (min)	(nm)	(eV)	
As deposited: Cu_2O	70	$Cu(S_2O_3)^-$	NaOH	10	20	60	14	2.20	[177]
200°C: Cu_2O							14	2.20	
250°C: Cu_2O							14	2.20	
300 °C: CuO							14–26	1.35	
350°C: CuO							–	1.35	
As deposited: ZnO	90	$ZnCl_2$	H_2O	150	10	60	18.4	3.1–3.2	[176]
200°C: ZnO							23.1		
300°C: ZnO							26.9		
400°C: ZnO							27.6		
As deposited: CuO	—	$CuCl_2$	NH_3	80	30	30	11.09	1.17	[178]
200°C: CuO							12.05	1.29	
300°C: CuO							13.86	1.30	
400°C: CuO							14.88	1.36	
As deposited: CuS	RT	$CuCl_2$	Na_2S	40	30	3 (N_2)	16.4	2.03	[43]
200°C: CuS							13.1	1.99	
300°C: CuS							33.6	1.97	
400°C: CuS							37.6	1.87	
As deposited: $Cu_{0.6}Zn_{0.4}S$	RT	$CuCl_2 + ZnCl_2$	Na_2S	40	30	3 (N_2)	38.6	2.14	[43]
200°C: $Cu_{0.6}Zn_{0.4}S$							29.9	2.12	
300°C: $Cu_{0.6}Zn_{0.4}S$							30.8	2.05	
400°C: $Cu_{0.6}Zn_{0.4}S$							27.9	1.85	
As deposited: ZnS	RT	$ZnCl_2$	Na_2S	40	30	3 (N_2)	2.7	3.92	[43]
200°C: ZnS							1.9	3.84	
300°C: ZnS							3.7	3.67	
400°C: ZnS							6.3	3.47	
250°C: CdO	70	$Cd(NO_3)_2$	H_2O_2	40	20	120	6.11	2.44	[179]
350°C: CdO							8.18	2.37	
450°C: CdO							9.48	2.24	

Temperature, ^{0}C		Precursors		Cycle	Time		Crystallite	Bandgap	Ref
Annealing	Growth	Cationic	Anionic		Dipping (sec)	Annealing (min)	(nm)	(eV)	
As deposited: MgO	90	$MgSO_4$	H_2O	30	30	60	12	4.18	[180]
100°C: MgO							10	4.18	
200°C: MgO							10	4.22	
300°C: MgO							10	4.25	
400°C: MgO							6	4.27	
500°C: MgO							6	4.33	
As deposited: NiO	90	$NiCl_2$	H_2O	100	30	30	10.03	3.3	[31]
200°C: NiO							11.04	3.25	
300°C: NiO							8.24	3.21	
400°C: NiO							13.32	3.11	

Table 8.
A list of thin films grown via *SILAR and annealed further for better film quality.*

Product	Precursors		Dopant		No of	Crystallite	Grain	Bandgap	Ref
	Cationic	Anionic	Compound	Amount	Cycle	(nm)	(nm)	(eV)	
Mn: Fe_2O_3	$FeCl_2$	NaOH	$MnCl_2$ (wt %)	0	50	28	—	3.02	[187]
				2		26		3.01	
				4		23		3.00	
				6		21		2.97	
				8		20		2.95	
Cu: Mn_3O_4	$Mn(NO_3)_2$	NH_4OH	$Cu(NO_3)_2$ (at %)	0	20	29	—	2.09	[188]
				0.1		25.7		2.06	
				0.2		28.4		1.91	
				0.3		30		1.71	
Co: CuO	$CuCl_2$	NH_4OH	$CoCl_2$ (at %)	0	10	22.7	70	1.53	[189]
				0.5		15.7	44	1.47	
				1		13.6	42	1.45	
				2		13.1	36	1.41	
				3		12.6	32	1.38	
				4		12.2	38	1.36	
B: CuO	$CuCl_2$	NH_4OH	H_3BO_4 (at %)	0	10	12.9	45	1.52	[190]
				1		13.1	42	1.48	
				2		14.2	38	1.43	
				3		15.9	30	1.39	
Co: Cu_2O	$Cu(S_2O_3)^-$	NaOH	$CoSO_4$ (wt %)	0	30	62.83	—	1.94	[191]
				1		53.30		2.03	
				2		48.47		2.12	
				5		39.24		2.18	
				10		28.44		2.47	
Fe: Cu_2O	$Cu(S_2O_3)^-$	NaOH	$FeSO_4$ (wt %)	0	30	62.83	—	1.80	[192]
				1		59.80		2.10	
				2		41.83		2.36	
				5		36.40		2.45	
Cr: CdO	$Cd(NO_3)_2$	NH_4OH	$Cr(NO_3)_3$ (M %)	0	20	129	1000	2.50	[193]
				0.1		121	1000	2.63	

Product	Precursors		Dopant		No of	Crystallite	Grain	Bandgap	Ref
	Cationic	Anionic	Compound	Amount	Cycle	(nm)	(nm)	(eV)	
				0.3		129	1000	2.78	
				0.5		131	1000	2.87	
Mn: CdO	$Cd(CH_3COO)_2$	NH_4OH	Mn solution (M %)	0	15	33.40	50	2.23	[194]
				1		39.26	to	2.29	
				2		38.15	35	2.39	
Ag: CdO	$Cd(CH_3COO)_2$	NH_4OH	Ag^+ solution (M %)	0	10	21	—	2.12	[195]
				1		15		2.34	
				2		17		2.27	
				3		19		2.18	
Mo: CdS	$Cd(CH_3COO_2)$	$SC(NH_2)_2$	$(NH_4)_6Mo_7O_{24}$ (at %)	0	70	—	~50	2.45	[196]
				5			~50	2.46	
				10			~50	2.47	
				15			~50	2.48	
Sn: CdS	$Cd(CH_3COO)_2$	CH_4N_2S	$SnCl_2$ (at %)	0	70	27.20	—	2.48	[197]
				2		32.79		2.50	
				4		24.01		2.51	
				6		18.87		2.51	
				8		12.96		2.43	
Ba: PbS	$Pb(CH_3COO)_2$	CH_3CSNH_2	$BaCl_2$ (at %)	0	200	18.6	480	1.77	[198]
				1		24.8	570	1.86	
				2		22.4	650	1.98	
				4		26.8	475	2.00	
				6		27.6	840	2.08	
				8		22.3	930	2.12	
Co: ZnS	$Zn(NO_3)_2$	Na_2S	Undoped	0	15	34	60	4.12	[199]
			$Cu(NO_3)_2$	3%		41	to	3.99	
			$Co(NO_3)_2$	3%		49	90	3.90	
			$AgNO_3$	3%		57		3.78	
F: ZnO	$Zn(CH_3COO)_2$	NaOH	NH_4F (at %)	0	50	49.67	—	3.21	[200]
				5		83.22		3.24	

Product	Precursors		Dopant		No of	Crystallite	Grain	Bandgap	Ref
	Cationic	Anionic	Compound	Amount	Cycle	(nm)	(nm)	(eV)	
				10		52.01		3.23	
				15		48.98		3.20	
Mg: ZnO	$ZnSO_4$	NaOH	$Cu(CH_3COO)_2$ (wt %)	0	50	37.6	—	3.09	[201]
				1		36.6		3.16	
				3		36.2		3.24	
				5		35.1		3.35	

Table 9.
A list of the doped films with some of their properties, fabricated by using SILAR techniques.

valence band to conduction band and doping is one of the best options in such structure tuning [181, 182]. In many cases, the incorporation of cation doping is an effective way to improve the electrical conductivity [183, 184]. By decreasing the bandgap, electron transfer between the valence and conduction bands will increase, and thus in case of energy storage device, electrode capacitance will increase [185]. For example, Y-doping in $Sr.(OH)_2$ improves both electronic conductivity as well as electrochemical performance of the electrode for energy storage device [89].

Again, electrochemical performance of In^{3+}-doped $WO_3/[Cu_7Te_4/Bi_2Te_3]$ electrodes for similar applications enhanced the capacitance to a great extent [90]. The study showed the specific capacity of undoped WO_3 was around 64 $mAhg^{-1}$ whereas it was increased to 90.2 $mAhg^{-1}$ for the same scan rate of 10 mVs^{-1} for In^{3+}-doped films, with the high-power density of 1.7 $kWkg^{-1}$ at the highest energy density of 18.85 $Whkg^{-1}$. Inside the WO_3 lattice, the doped In^{3+} cation diffused and was connected in the insertion-removal exchange method of electrons, with further electrochemical S^{2-} insertion or extraction striking at the Cu_7Te_4/Bi_2Te_3 and polysulfide electrolyte surfaces.

Besides, Zhu and coauthors prepared Cu-doped CdS on QDSCs and investigated the effect of Cu doping on several cells based on the doping concentration. When the doping ratio of Cu decreased successively from 1:10, all the parameters such as J_{sc}, V_{oc}, and PEC increased and reached at the maximum value when the ratio became 1: 500 [186]. In a separate report of Mn-doped-CdS/CdSe deposited on mesoscopic TiO_2 film as photoanode using Cu2S/graphene oxide composite electrode, in the presence of sulfide/polysulfide electrolyte provide PEC of 5.4%, higher than undoped sample [118], whereas Hg^{2+}-doped PbS QDSC having unprecedentedly high photocurrent delivered PEC of 5.6% at one sun illumination over the undoped PbS QD cell [120]. Moreover, Abel and co-workers developed an improved photoelectrochemical water splitting device *via* SILAR-fabricated Ti-doped α-Fe_2O_3 thin films. Ti, acting as water oxidation intermediates, enhanced interfacial hole transfer efficiency from less than 3–80% by increasing the concentration of surface-trapped holes, which is then triggered by FeOOH to amplify hole transfer efficiency to ~100%. Both Ti doping and FeOOH overlayer resulted in photocurrents of 0.85 $mAcm^{-2}$ at 1.23 V vs. RHE [138]. However, a lot of work has been done by several authors to get better film quality through doping. A list of such initiatives of the doped films with the starting materials of growth and other properties is shown in **Table 9**.

6. Conclusion

This chapter represented detailed discussions on the methods and techniques of the fabrication process of thin films by utilizing SILAR for the optoelectronic device applications. Among the diverse fabrication techniques both physically and chemically, SILAR is the simplest to fabricate thin films having remarkable quality. It is widely fit for the fabrication of thin films of metal chalcogenides, hydroxides, peroxides, as well as complex and composite nanostructures with innovative functionalities. The role of experimental conditions on the structural, optical, and electrical properties of the thin films as well as device performances is reviewed in this chapter mainly for the advanced utilization of both the generation and storage of energy such as solar cells, photoelectrochemical water splitting, supercapacitors, and so on. The technological advancement of a fabrication technique is deeply reliant on the opportunity of controlling the experimental factors involved. In this chapter, a brief advantage of SILAR technique is highlighted, including flexibility of the film growth, thickness control, composition control, and low temperature management, along with a broad range of applications. From this point of view, a deep knowledge of the connections between processing, structure, specific characteristics, and performances is the foundation for accurate and rational engineering of such optoelectronic devices. Moreover, a comprehensive profile of recent status is required to focus on further prospects. This work will therefore deliver a strong contribution to move ahead with future research goals on SILAR technique, by utilizing low-cost deposition of high-quality thin films and associated optoelectronic devices.

Acknowledgements

The author thanks the Department of Chemistry, Comilla University and UGC, Bangladesh, for supporting this work. Also, thanks to Professor Dr. Jamal Uddin, Center for Nanotechnology, Department of Natural Sciences, Coppin State University, Baltimore, MD, USA, for his insightful discussion and support.

Conflict of interest

The authors declare no conflict of interest.

Author details

Md Abdul Majed Patwary
Department of Chemistry, Comilla University, Cumilla, Bangladesh

*Address all correspondence to: mamajedp@gmail.com

References

[1] Nicolau Y. Solution deposition of thin solid compound films by a successive ionic-layer adsorption and reaction process. Applied Surface Science. 1985; **22**:1061

[2] Ristov M, Sinadinovski G, Grozdanov I. Chemical deposition of Cu2O thin films. Thin Solid Films. 1985;**123**:63

[3] Pathan H, Lokhande C. Deposition of metal chalcogenide thin films by successive ionic layer adsorption and reaction (SILAR) method. Bulletin of Materials Science. 2004;**27**: 85-111

[4] Nicolau Y, Dupuy M, Brunel M. ZnS, CdS, and $Zn_{1-x}Cd_xS$ Thin Films Deposited by the Successive Ionic Layer Adsorption and Reaction Process. Journal of the Electrochemical Society. 1990;**137**:2915

[5] Nicolau Y, Menard J. Solution growth of ZnS, CdS and Zn_{1-X} Cd_X S thin films by the successive ionic-layer adsorption and reaction process; growth mechanism. Journal of Crystal Growth. 1988;**92**:128

[6] Hossain MA, Farhad SFU, Tanvir NI, Chang JH, Rahman MA, Tanaka T, et al. Facile synthesis of Cu2O nanorods in the presence of NaCl by successive ionic layer adsorption and reaction method and its characterizations. RSC Open Science. 2022;**9**:211899

[7] Ghos BC, Farhad SFU, Patwary MAM, Majumder S, Hossain MA, Tanvir NI, et al. Influence of the Substrate, Process Conditions and Post-annealing Temperature on the Properties of ZnO Thin Films Grown by the Successive Ionic Layer Adsorption and Reaction Method. Journal of ACS Omega. 2021;**6** (4):2665

[8] Majumder S, Tanvir NI, Ghos BC, Patwary MAM, Rahman MA, Hossain MA, et al. Optimization of the growth conditions of Cu_2O thin films and subsequent fabrication of Cu_2O/ZnO heterojunction by m-SILAR method. IEEE WIECON-ECE. 2020:139

[9] Ubale AU, Belkhedkar MR, Sakhare YS, Singh A, Gurada C, Kothari DC. Characterization of nanostructured Mn3O4 thin films grown by SILAR method at room temperature. Materials Chemistry and Physics. 2012;**136**:1067

[10] Xiangdong G, Xiaomin L, Weidong Y, Preparation and characterization of highly oriented ZnO film by ultrasonic assisted SILAR method. Journal of Wuhan University of Technology, Materials Science Edition. 2005;**20**:23

[11] Sfaelou S, Sygellou L, Dracopoulos V, Travlos A, Lianos P. Effect of the nature of Cadmium salts on the effectiveness of CdS SILAR deposition and Its consequences on the Performance of Sensitized Solar Cells. Journal of Physical Chemistry C. 2014; **118**:22873

[12] Wang T, Luo Z, Li C, Gong J. Controllable fabrication of nanostructured materials for photoelectrochemical water splitting via atomic layer deposition. Chemical Society Reviews. 2014;**43**:7469

[13] Pawar SM, Pawar BS, Kim JH, Joo O-S, Lokhande CD. Recent status of chemical bath deposited metal chalcogenide and metal oxide thin films. Current Applied Physics. 2011;**11**:117

[14] Niesen TP, De Guire MR. Review: Deposition of Ceramic Thin Films at Low Temperatures from Aqueous

Solution. Journal of Electroceramics. 2001;**6**:169

[15] Sankapal BR, Ennaoui A, Guminskaya T, Dittrich T, Bohne W, Röhrich J, et al. Characterization of p-CuI prepared by the SILAR technique on Cu-tape/n-CuInS_2 for solar cells. Thin Solid Film. 2005;**480-481**:142-146

[16] Oluyamo SS, Nyagba SM, Ojo SA. Optical Properties of Copper (I) Oxide Thin Films Synthesized by SILAR Technique. IOSR Journal of Applied Physics. 2014;**6**:102-110

[17] Mageshwari K, Sathyamoorthy R. Physical properties of nanocrystalline CuO thin films prepared by the SILAR method. Materials Science in Semiconductor Processing. 2012;**16**:337-343

[18] Mehrabian M. Optical and photovoltaic properties of ZnS nanocrystals fabricated on Al:ZnO films using the SILAR technique. Journal of Optical Technology. 2016;**83**:422-428

[19] García C, Dávila S, Jardón G, Flores F, Bon R, Vorobiev YV. Characterization of PbS films deposited by successive ionic layer adsorption and reaction (SILAR) for CdS/PbS solar cells application. Materials Research Express. 2020;7:015530

[20] Manikandan K, Dilip C, Mani P, Prince JJ. Deposition and Characterization of CdS Nano Thin Film with Complexing Agent Triethanolamine. American Journal of Engineering & Applied Science. 2015;**8**:318

[21] Lokhande C, Shelke H, Raut V, Patil A, Lokhande A, Kim J. Facile synthesis of Cu2SnS3 thin films grown by SILAR method: effect of film thickness. Journal of Materials Science: Materials in Electronics. 2017;**28**:7912

[22] Hodes G. Chemical Solution Deposition of Semiconductor Films. New York: Marcel Dekker; 2005

[23] Lundin A, Kitaev G. Inorganic Materials. 1965;**1**:1900-1905

[24] Nikam S, Suryawanshi M, Bhosale S, Gaikwad M, Shinde P, Moholkar A. Cu_2O thin films prepared using modified successive ionic layer adsorption and reaction method and their use in photoelectrochemical solar cells. Journal of Materials Science: Materials in Electronics. 1897;**2016**:27

[25] Das MR, Mukherjee A, Maiti P, Das S, Mitra P. Studies on multifunctional properties of SILAR synthesized CuO thin films for enhanced supercapacitor, photocatalytic and ethanol sensing applications. Journal of Electronic Materials. 2019;**48**:2718

[26] Garcia FJ, Calderon CLL, Arbelaez DE, Del Real A, Garcia MR. Influence of substrate on structural, morphological and optical properties of ZnO films grown by SILAR method. Bulletin of Materials Science. 2014;**37**:1283

[27] Gaikwad M, Mane A, Desai S, Moholkar A. Template-free TiO_2 photoanodes for dye-sensitized solar cell via modified chemical route. Journal of Colloid Interface, Science. 2017;**488**:269

[28] Jambure S, Lokhande C. Photoelectrochemical solar cells with chemically grown CdO rice grains on flexible stainless-steel substrates. Materials Letters. 2013;**106**:133

[29] Thakur AV, Lokhande BJ. Electrolytic anion affected charge storage mechanisms of Fe_3O_4 flexible thin film electrode in KCl and KOH: a comparative study by cyclic voltammetry and galvanostatic charge-

discharge. Journal of Materials Science: Materials in Electronics. 2017;**28**:11755

[30] Dubal DP, Jagadale AD, Lokhande CD. Big as well as light weight portable, Mn3O4 based symmetric supercapacitive devices: Fabrication, performance evaluation and demonstration. Electrochimica Acta. 2012;**80**:160

[31] Taşdemirci TCA. Influence of Annealing on Properties of SILAR Deposited Nickel Oxide Films. Vacuum. 2019;**167**:189

[32] Shinde NM, Jagadale AD, Kumbhar VS, Rana TR, Kim J, Lokhande CD. Wet chemical synthesis of WO3 thin films for supercapacitor application. Korean Journal of Chemical Engineering. 2015; **32**:974

[33] Raut SS, Bisen O, Sankapal BR. Synthesis of interconnected needle-like Bi2O3 using successive ionic layer adsorption and reaction towards supercapacitor application. Ionics. 1831;**2017**:23

[34] Guney H, İskenderoğlu D. Synthesis of MgO thin films grown by SILAR technique, Synthesis of MgO thin films grown by SILAR technique. Ceramics International. 2018;**44**:7788

[35] Guo W, Lian X, Nie Y, Hu M, Wu L, Gao H, et al. Facile growth of b-Cu2V2O7 thin films and characterization for photoelectrochemical water oxidation. Materials Letters. 2020;**258**: 126842

[36] Chemelewski WD, Mabayoje O, Mullins CB. SILAR Growth of Ag3VO4 and Characterization for Photoelectrochemical Water Oxidation. Journal of Physical Chemistry C. 2015; **119**:26803

[37] Guo W, Tang D, Mabayoje O, Wygant BR, Xiao P, Zhang Y, et al. A Simplified Successive Ionic Layer Adsorption and Reaction (s-SILAR) Method for Growth of Porous BiVO4 Thin Films for Photoelectrochemical Water Oxidation. Journal of the Electrochemical Society. 2017;**164**:H119

[38] Tang D, Rettie AJ, Mabayoje O, Wygant BR, Lai Y, Liu Y, et al. Facile Growth of Porous Fe_2V4O_{13} Films for Photoelectrochemical Water Oxidation. Journal of Materials Chemistry A. 2016; **4**:3034

[39] Vadiyar MM, Kolekar SS, Deshpande NG, Chang J-Y, Kashale AA, Ghule AV. Binder-free chemical synthesis of $ZnFe_2O_4$ thin films for asymmetric supercapacitor with improved performance. Ionics. 2017;**23**:741

[40] Gong J, Wang X, Li X, Wang K. Highly sensitive visible light activated photoelectrochemical biosensing of organophosphate pesticide using biofunctional crossed bismuth oxyiodide flake arrays. Biosensors & Bioelectronics. 2012;**38**:43

[41] Chavan HS, Hou B, Ahmed ATA, Jo Y, Cho S, Kim J, et al. Nanoflake $NiMoO_4$ based smart supercapacitor for intelligent power balance monitoring. Solar Energy Materials & Solar Cells. 2018;**185**:166

[42] Babar P, Lokhande A, Shim H, Gang M, Pawar B, Pawar S, et al. SILAR deposited iron phosphate as a bifunctional electrocatalyst for efficient water splitting. Journal of Colloid and Interface Science. 2019;**534**:350

[43] Ali Yildirm M, Ates A, Astam A. Annealing and light effect on structural, optical and electrical properties of CuS, CuZnS and ZnS thin films grown by the

SILAR method. Physica E. 2009;**41**:1365-1372

[44] Paul G, Chatterjee S, Pal AJ. Heterovalent doping and energy level tuning in Ag_2S thin-films through solution approach: pn-Junction solar cells. Solar Energy Materials & Solar Cells. 2018;**182**:339

[45] Pathan H, Salunkhe P, Sankapal B, Lokhande C. Photoelectrochemical investigation of Ag2S thin films deposited by SILAR method. Materials Chemistry and Physics. 2001;**72**:105

[46] Veerathangam K, Pandian MS, Ramasamy P. Size-dependent photovoltaic performance of cadmium sulfide (CdS) quantum dots for solar cell applications. Journal of Alloys and Compounds. 2018;**735**:202

[47] Sartale S, Lokhande C. Preparation and characterization of nickel sulphide thin films using successive ionic layer adsorption and reaction (SILAR) method. Materials Chemistry and Physics. 2001;**72**:101

[48] Gao C, Shen H, Sun L, Huang H, Lu L, Cai H. Preparation of SnS films with zinc blende structure by successive ionic layer adsorption and reaction method. Materials Letters. 2010;**64**:2177

[49] Abbas MA, Basit MA, Park TJ, Bang JH. Enhanced performance of PbS-sensitized solar cells via controlled successive ionic-layer adsorption and reaction. Physical Chemistry Chemical Physics. 2015;**17**:9752

[50] Sartale S, Lokhande C. Studies on large area (~50 cm2) MoS2 thin films deposited using successive ionic layer adsorption and reaction (SILAR) method. Materials Chemistry and Physics. 2001;**71**:94

[51] Sartale S, Lokhande C. Preparation and characterization of As_2S_3 thin films deposited using successive ionic layer adsorption and reaction (SILAR) method. Materials Research Bulletin. 2000;**35**:1345

[52] Wang Y, Chen J, Jiang L, Liu F, Lai Y, Li J. Characterization of Bi_2S_3 thin films synthesized by an improved successive ionic layer adsorption and reaction (SILAR) method. Materials Letters. 2017;**209**:479

[53] Yıldırım MA, Yıldırım ST, Cavanmirza İ, Ateş A. Chemically synthesis and characterization of MnS thin films by SILAR method. Chemical Physics Letters. 2016;**647**:73

[54] Deshpande M, Chauhan K, Patel KN, Rajput P, Bhoi HR, Chaki S. Study of Sb2S3 thin films deposited by SILAR method. Materials Research Express. 2018;**5**:056410

[55] Karade SS, Dwivedi P, Majumder S, Pandit B, Sankapal BR. First report on FeS based 2 V operating flexible solid-state symmetric supercapacitor device. Sustainable Energy & Fuels. 2017;**1**:1366

[56] Manikandan K, Mani P, Dilip CS, Valli S, Inbaraj PFH, Prince JJ. Effect of complexing agent TEA: The structural, morphological, topographical and optical properties of FexSx nano thin films deposited by SILAR technique. Applied Surface Science. 2014;**288**:76

[57] Sartale S, Lokhande C. Deposition of cobalt sulfide thin films by successive ionic layer adsorption and reaction (SILAR) method, and their characterization. Indian Journal of Pure and Applied Physics. 2000;**38**:48

[58] Patil S, Lokhande A, Lokhande C. Effect of aqueous electrolyte on pseudocapacitive behavior of chemically

synthesized La2S3 electrode. Materials Science in Semiconductor Processing. 2016;**41**:132

[59] Astam A, Akaltun Y, Yıldırım M. Conversion of SILAR deposited Cu_3Se_2 thin films to $Cu_{2-x}Se$ by annealing. Materials Letters. 2016;**166**:9

[60] Zhao B, Wan Z, Luo J, Han F, Malik HA, Jia C, et al. Efficient Sb_2Se_3 sensitized solar cells prepared through a facile SILAR process and improved performance by interface modification. Applied Surface Science. 2018;**450**:228

[61] Chaudhari K, Gosavi NM, Deshpande N, Gosavi S. Chemical synthesis and characterization of CdSe thin films deposited by SILAR technique for optoelectronic applications. Journal of Science: Advanced Materials Devices. 2016;**1**:476

[62] Lokhande C, Sankapal B, Sartale S, Pathan H, Giersig M, Ganesan V. A novel method for the deposition of nanocrystalline Bi_2Se_3, Sb_2Se_3 and Bi_2Se_3–Sb_2Se_3 thin films — SILAR. Applied Surface Science. 2001;**182**:413

[63] Ubale A, Dhokne R, Chikhlikar P, Sangawar V, Kulkarni D. Characterization of nanocrystalline cadmium telluride thin films grown by successive ionic layer adsorption and reaction (SILAR) method. Bulletin of Materials Science. 2006;**29**:165

[64] Pathan H, Lokhande C, Amalnerkar D, Seth T. Preparation and characterization of copper telluride thin films by modified chemical bath deposition (M-CBD) method. Applied Surface Science. 2003;**218**:291

[65] Patil S, Lokhande A, Lee D-W, Kim J, Lokhande C. Chemical synthesis and supercapacitive properties of lanthanum telluride thin film. Journal of Colloid and Interface Science. 2017;**490**:147

[66] Kulal PM, Dubal DP, Lokhande CD, Fulari VJ. Chemical synthesis of Fe2O3 thin films for supercapacitor application. Journal of Alloys and Compounds. 2011; **509**:2567

[67] Yıldırım MA, Akaltun Y, Ateş A. Characteristics of SnO_2 thin films prepared by SILAR, Characteristics of SnO_2 thin films prepared by SILAR. Solid State Sciences. 2012;**14**:1282

[68] Yassin SNH, Sim ASL, Jennings JR. Photoelectrochemical evaluation of SILAR-deposited nanoporous BiVO4 photoanodes for solar-driven water splitting. Nano Materials Science. 2020; **2**:227

[69] Putri AA, Kato S, Kishi N, Soga T. Relevance of precursor molarity in the prepared bismuth oxyiodide films by successive ionic layer adsorption and reaction for solar cell application. Journal of Science: Advanced Materials Devices. 2019;**4**:116

[70] Sfaelou S, Raptis D, Dracopoulos V, Lianos P. BiOI solar cells. RSC Advances. 2015;**5**:95813

[71] Hu L, Liao Y, Xia D, Zhang Q, He H, Yang J, et al. In-situ fabrication of AgI-BiOI nanoflake arrays film photoelectrode for efficient wastewater treatment, electricity production and enhanced recovery of copper in photocatalytic fuel cell. Catalysis Today. 2020;**339**:379

[72] Zhang Y, Chong X, Sun H, Kedir MM, Kim KJ, Ohodnicki PR, et al. Silver nanoplate aggregation based multifunctional black metal absorbers for localization, photothermic harnessing enhancement and omnidirectional light antireflection.

Journal of Materials Chemistry C. 2018; **6**:989

[73] Mehrabian M, Esteki Z. Degradation of methylene blue by photocatalysis of copper assisted ZnS nanoparticle thin films. Optik. 2017;**130**:1168

[74] Zhao J, Burke AF. Electrochemical Capacitors: Performance Metrics and Evaluation by Testing and Analysis. Advanced Energy Materials. 2020;**11**: 2002192

[75] Dubal D, Dhawale D, Salunkhe R, Lokhande C. A novel chemical synthesis of Mn3O4 thin film and its stepwise conversion into birnessite MnO2 during super capacitive studies. Journal of Electroanalytical Chemistry. 2010;**647**:60

[76] Nwankwo M, Nwanya A, Agbogu A, Ekwealor A, Ejikeme PM, Bucher R, et al. Electrochemical Supercapacitive Properties of SILAR-Deposited Mn_3O_4 Electrodes. Vacuum. 2018;**158**:206

[77] Kumbhar VS, Lee YR, Ra CS, Tuma D, Min B-K, Shim JJ. Modified chemical synthesis of MnS nanoclusters on nickel foam for high performance all-solid-state asymmetric supercapacitors. RSC Advances. 2017;7:16348

[78] Das MR, Roy A, Mpelane S, Mukherjee A, Mitra P, Das S. Influence of dipping cycle on SILAR synthesized NiO thin film for improved electrochemical performance. Electrochimica Acta. 2018;**273**:105

[79] Shinde S, Ghodake G, Fulari V, Kim D-Y. High electrochemical performance of nanoflakes like CuO electrode by successive ionic layer adsorption and reaction (SILAR) method. Journal of Industrial and Engineering Chemistry. 2017;**52**:12

[80] Shinde SK, Yadav HM, Ramesh S, Bathula C, Maile N, Ghodake GS, et al. High -Performance Symmetric Supercapacitor; Nanoflower -Like $NiCo_2O_4//NiCo_2O_4$ Thin Films Synthesized by simple and highly stable chemical method. Journal of Molecular Liquids. 2020;**299**:112119

[81] Chavan HS, Hou B, Ahmed ATA, Kim J, Jo Y, Cho S, et al. Ultrathin Ni-Mo oxide nanoflakes for high-performance supercapacitor electrodes. Journal of Alloys and Compounds. 2018;**767**:782

[82] Pusawale SN, Deshmukh PR, Jadhav PS, Lokhande CD. Electrochemical properties of chemically synthesized SnO_2-RuO_2 mixed films. Materials for Renewable and Sustainable Energy. 2018;**8**:1

[83] Kumar N, Mishra D, Kim SY, Jin SH. Two dimensional, bi-layered SnS_2@Co_3S_4 heterostructure formation via SILAR method: Toward high performance supercapacitors with superior electrodes. Materials Letters. 2020;**262**:127173

[84] Gund GS, Dubal DP, Shinde SS, Lokhande CD. Architectured Morphologies of Chemically Prepared NiO/MWCNTs Nanohybrid Thin Films for High Performance Supercapacitors. ACS Applied Materials & Interfaces. 2014;**6**:3176

[85] Kumar N, Sahoo P, Panda H. Tuning the electro-chemical properties by substituting selectively transition metals on carbon in Ni/Co oxide-carbon composite electrode for supercapacitor device. New Journal of Chemistry. 2017; **41**:3562

[86] Pandit B, Pande SA, Sankapal BR. Facile SILAR Processed Bi_2S_3:PbS Solid Solution on MWCNTs for High-performance Electrochemical

Supercapacitor. Chinese Journal of Chemistry. 2019;**37**:1279

[87] Raut SS, Sankapal BR. Porous zinc cobaltite (ZnCo2O4) film by successive ionic layer adsorption and reaction towards solid-state symmetric supercapacitive device. Journal of Colloid and Interface Science. 2017;**487**: 201

[88] Raut SS, Sankapal BR. First report on synthesis of $ZnFe_2O_4$ thin film using successive ionic layer adsorption and reaction: Approach towards solid-state symmetric supercapacitor device. Electrochimica Acta. 2016;**198**:203

[89] Kavyashree S, Parveen SK, Sharma SN. Solid-state symmetric supercapacitor based on Y doped Sr $(OH)_2$ using SILAR method. Pandey, Energy. 2020;**197**:117163

[90] Buathet S, Simalaotao K, Reunchan P, Vailikhit V, Teesetsopon P, Raknual D, et al. Tubtimtae, Electrochemical performance of Bi_2Te_3 heterostructure thin film and Cu_7Te_4 nanocrystals on undoped and In^{3+}-doped WO_3 films for energy storage applications. Electrochimica Acta. 2020;**341**:136049

[91] Becquerel E. On Electron Effects under the Influence of Solar Radiation. Comptes. 1839;**9**(144):561

[92] Fritts CE. On a New Form of Selenium Cell, and some Electrical Discoveries made by its use. American Journal of Science. 1883;**26**:465

[93] Ohl R. U. S. Patent 2. 1946;**402**:662

[94] Chapin D, Fuller C, Pearson G. A New silicon p-n Junction Photocell for Converting Solar Radiation into Electrical Power. Journal of Applied Physics. 1954;**25**(5):676

[95] National Renewable Energy Laboratory, NREL, USA. Available from: https://www.nrel.gov/pv/interactive-cell-efficiency.html

[96] Shockley W, Queisser HJ. Detailed Balance Limit of Efficiency of *p-n* Junction Solar Cells. Journal of Applied Physics. 1961;**32**(3):510

[97] Maycock PD. PV review: World Solar PV market continues explosive growth. Refocus. 2005;**6**:18

[98] Inaba T, Kobayashi E, Suwazono Y, Uetani M, Oishi M, Nakagawa H, et al. Estimation of Cumulative Cadmium Intake Causing Itai-itai Disease. Toxicology Letters. 2005;**159**:192-201

[99] Baba H, Tsuneyama K, Yazaki M, Nagata K, Minamisaka T, Tsuda T, et al. The liver in itai-itai disease (chronic cadmium poisoning): pathological features and metallothionein expression. Modern Pathology. 2013;**26**:1228

[100] Pan J, Plant JA, Voulvoulis N, Oates CJ, Ihlenfeld C. Ihlenfeld, Cadmium levels in Europe: implications for human health. Environmental Geochemistry and Health. 2010; **32**:1-12

[101] Bertin G, Averbeck D. Cadmium: cellular effects, modifications of biomolecules, modulation of DNA repair and genotoxic consequences (a review). Biochimie. 2006;**88**:1549-1559

[102] Gaikwad M, Suryawanshi M, Maldar P, Dongale T, Moholkar A. Nanostructured zinc oxide photoelectrodes by green routes M-SILAR and electrodeposition for dye sensitized solar cell. Optical Materials. 2018;**78**:325

[103] Divitini G, Cacovich S, Matteocci F, Cinà L, Di Carlo A, Ducati C. *In situ*

observation of heat-induced degradation of perovskite solar cells. Nature Energy. 2016;**1**:15012

[104] Pathak SK, Abate A, Leijtens T, Hollman DJ, Teuscher J, Pazos L, et al. Towards Long-Term Photostability of Solid-State Dye Sensitized Solar Cells. Advanced Energy Materials. 2014;**4**(8): 1301667

[105] Rühle S, Anderson AY, Barad HN, Kupfer B, Bouhadana Y, Hodesh ER, et al. All-Oxide Photovoltaics. Journal of Physical Chemistry Letters. 2012;**3**:3755

[106] Dittrich T, Belaidi A, Ennaoui A. Ennaoui, Concepts of inorganic solid-state nanostructured solar cells. Solar Energy Materials & Solar Cells. 2011;**95**: 1527

[107] Thakur AV, Lokhande BJ. Effect of the molar concentration of pyrrole monomer on the rate of polymerization, growth and hence the electrochemical behavior of highly pristine PPy flexible electrodes. Heliyon. 2019;**5**:e02909

[108] Badawi A, Al-Baradi AM, Atta AA, Algarni SA, Almalki ASA, Alharthi SS. Graphene/TiO_2 nanocomposite electrodes sensitized with tin sulfide quantum dots for energy issues. Physica E. 2020;**121**:114121

[109] Wang QH, Kalantar-zadeh K, Coleman KAJN, Strano MS. Electronics and optoelectronics of two-dimensional transition metal dichalcogenides. Nature Nanotechnology. 2012;**7**:699-712

[110] Manzeli S, Ovchinnikov D, Pasquier D, Yazyev OV, Kis A. 2D transition metal dichalcogenides. Nature Reviews Materials. 2017;**2**(8):17033

[111] Tack LW, Azam MA, Seman RNAR. Structural and Electronic Properties of Transition-Metal Oxides Attached to a Single-Walled CNT as a Lithium-Ion Battery Electrode: A First-Principles Study. The Journal of Physical Chemistry A. 2017;**121**(13):2636

[112] Chatterjee S, Pal AJ. Introducing Cu2O Thin-Films as a Hole-Transport Layer in Efficient Planar Perovskite Solar Cell Structures. Journal of Physical Chemistry C. 2016;**120**:1428

[113] Gaikwad M, Suryawanshi M, Nikam S, Bhosale C, Kim J, Moholkar A. Influence of Zn concentration and dye adsorption time on the photovoltaic performance of M- SILAR deposited ZnO-based dye sensitized solar cells. Journal of Photochemistry and Photobiology, A: Chemistry. 2016;**329**:246

[114] Shaikh SF, Ghule BG, Nakate UT, Shinde PV, Ekar SU, O'Dwyer C, et al. Low-Temperature Ionic Layer Adsorption and Reaction Grown Anatase TiO2 Nanocrystalline Films for Efficient Perovskite Solar Cell and Gas Sensor Applications. Scientific Reports. 2018;**8**: 11016

[115] Sankapal B, Lokhande CD. Photoelectrochemical characterization of Bi2Se3 thin films deposited by SILAR technique. Materials Chemistry and Physics. 2002;**73**:151

[116] Luo J, Wang YX, Sun J, Yang ZS, Zhang QF. MnS passivation layer for highly efficient ZnO-based quantum dotsensitized solar cells. Solar Energy Materials & Solar Cells. 2018;**187**:199

[117] Becker MA, Radich JG, Bunker BA, Kamat PV. How Does a SILAR CdSe Film Grow? Tuning the Deposition Steps to Suppress Interfacial Charge Recombination in Solar Cells. Journal of Physical Chemistry Letters. 2014;**5**:1575

[118] Santra PK, Kamat PV. Mn-Doped Quantum Dot Sensitized Solar Cells: A

Strategy to Boost Efficiency over 5%. Journal of the American Chemical Society. 2012;**134**:2508

[119] Zhou R, Zhang Q, Tian J, Myers D, Yin M, Cao G. Influence of Cationic Precursors on CdS Quantum-Dot-Sensitized Solar Cell Prepared by Successive Ionic Layer Adsorption and Reaction. Journal of Physical Chemistry C. 2013;**117**:26948

[120] Lee JW, Son D-Y, Ahn TK, Shin H-W, Kim IY, Hwang S-J, et al. Quantum-Dot-Sensitized Solar Cell with Unprecedentedly High Photocurrent. Scientific Reports. 2013;**3**:1050

[121] Tian J, Shen T, Liu X, Fei C, Lv L, Cao G. Enhanced Performance of PbS quantum-dot-sensitized Solar Cells via Optimizing Precursor Solution and Electrolytes. Scientific Reports. 2016;**6**:1

[122] Kumar DK, Loskot J, Křiž J, Bennett N, Upadhyaya HM, Sadhu V, et al. Synthesis of SnSe quantum dots by successive ionic layer adsorption and reaction (SILAR) method for efficient solar cells applications. Solar Energy. 2020;**199**:570

[123] Visalakshi S, Kannan R, Valanarasu S, Kim H-S, Kathalingam A, Chandramohan R. Effect of bath concentration on the growth and photovoltaic response of SILAR-deposited CuO thin films. Applied Physics A: Materials Science & Processing. 2015;**120**:1105

[124] Chatterjee S, Saha SK, Pal AJ. Formation of all-oxide solar cells in atmospheric condition based on Cu_2O thin films grown through SILAR technique. Solar Energy Materials & Solar Cells. 2016;**147**:17

[125] Jambure SB, Gund GS, Dubal DP, Shinde SS, Lokhande CD. Cost effective facile synthesis of TiO2 nanograins for flexible DSSC application using rose bengal dye. Electronic Materials Letters. 2014;**10**:943

[126] Waghmare M, Beedri N, Baviskar P, Pathan H, Ubale A. Effect of ZrO_2 barrier layers on the photovoltaic parameters of rose bengal dye sensitized TiO_2 solar cell. Journal of Materials Science: Materials in Electronics. 2019; **30**:6015

[127] Lee H, Wang M, Chen P, Gamelin DR, Zakeeruddin SM, Gratzel M, et al. Nazeeruddin, Efficient CdSe Quantum Dot-Sensitized Solar Cells Prepared by an Improved Successive Ionic Layer Adsorption and Reaction Process. Nano Letters. 2009;**9**:4221

[128] She H, Sun Y, Li S, Huang J, Wang L, Zhu G, et al. Synthesis of non-noble metal nickel doped sulfide solid solution for improved photocatalytic performance. Applied Catalysis, B: Environmental. 2019;**245**:439-447

[129] Zhou S, Yue P, Huang J, Wang L, She H, Wang Q. High-performance photoelectrochemical water splitting of BiVO4@Co-MIm prepared by a facile in-situ deposition method. Journal of Chemical Engineering. 2019;**371**:885-892

[130] Adam RE, Pirhashemi M, Elhag S, Liu X, Habibi-Yangjeh A, Willander M, et al. Nur, ZnO/Ag/Ag2WO4 photo-electrodes with plasmonic behavior for enhanced photoelectrochemical water oxidation. RSC Advances. 2019;**9**: 8271

[131] Pirhashemi M, Habibi-Yangjeh A. Ultrasonic-assisted preparation of plasmonic ZnO/Ag/Ag2WO4 nanocomposites with high visible-light photocatalytic performance for degradation of organic pollutants.

Journal of Colloid and Interface Science. 2017;**491**:216-229

[132] Ghobadifard M, Mohebbi S. Novel nanomagnetic Ag/β-Ag_2WO_4/$CoFe_2O_4$ as a highly efficient photocatalyst under visible light irradiation. New Journal of Chemistry. 2018;**42**:9530-9542

[133] Pan L, Zhao L, Liu Z. High-efficient TiO_2 NRs/BiOI NSs heterojunction photoanodes for photoelectrochemical water splitting,. Materials and Technologies. 2017;**32**:823

[134] Yassin BH, Halidi NAH, Sim SL, Liu YR, Jennings JR. Electron Diffusion Length and Charge Separation Efficiency in Nanostructured Ternary Metal Vanadate Photoelectrodes. Materials Science Forum. 2018;**941**:2121

[135] Xiong Y, Yang L, He H, Wan J, Xiao P, Guo W. Enhanced charge separation and transfer by Bi2MoO6@Bi2Mo2O9 compound using SILAR for photoelectrochemical water oxidation. Electrochimica Acta. 2018;**264**:26

[136] Stoll T, Zafeiropoulos G, Dogan I, Genuit H, Lavrijsen R, Koopmans B, et al. Tsampas, Visible-light-promoted gas-phase water splitting using porous WO_3/$BiVO_4$ photoanodes. Electrochemistry Communications. 2017;**82**:47

[137] Abel AJ, Garcia-Torregrosa I, Patel AM, Opasanont B, Baxter JB. SILAR-Deposited Hematite Films for Photoelectrochemical Water Splitting: Effects of Sn, Ti, Thickness, and Nano structuring. Journal of Physical Chemistry C. 2015;**119**:4454

[138] Abel AJ, Patel AM, Smolin SY, Opasanont B, Baxter JB. Enhanced Photoelectrochemical Water Splitting via SILAR-Deposited Ti-Doped Hematite Thin Films with an FeOOH Overlayer. Journal of Materials Chemistry A. 2016;**4**:6495

[139] Zhong X, He H, Du J, Ren Q, Huang J, Tang Y, et al. Boosting solar water oxidation activity and stability of $BiVO_4$ photoanode through the Co-catalytic effect of $CuCoO_2$. Electrochimica Acta. 2019;**304**:301

[140] Majumder S, Quang ND, Hien TT, Chinh ND, Hung NM, Yang H, et al. Effect of SILAR-anchored ZnFe2O4 on the BiVO4 nanostructure: An attempt towards enhancing photoelectrochemical water splitting. Applied Surface Science. 2021:149033

[141] Zhou M, Guo Z, Song Q, Li X, Liu Z. Improved photoelectrochemical response of $CuWO_4$/BiOI p-n heterojunction embedded with plasmonic Ag nanoparticles. Chemical Engineering Journal. 2019;**370**:218

[142] Xia L, Bai J, Li J, Zeng Q, Li L, Zhou B. High-performance BiVO4 photoanodes cocatalyzed with an ultrathin -Fe_2O_3 layer for photoelectrochemical application. Applied Catalysis, B: Environmental. 2017; **204**:127

[143] Zhou W, Jiang T, Zhao Y, Xu C, Pei C, Xue H. Ultrathin TiO2/BiVO4 nanosheet heterojunction arrays modified with NiFe-LDH nanoparticles for enhanced photoelectrochemical oxidation of water. Journal of Colloid and Interface Science. 2019;**549**:42

[144] Yan L, Zhao W, Liu Z. 1D ZnO/$BiVO_4$ heterojunction photoanodes for efficient photoelectrochemical water splitting, Dalton Transactions. 2016;**45**: 11346

[145] Gao L, Long X, Wei S, Wang C, Wang T, Li F, et al. Facile growth of $AgVO_3$ nanoparticles on Mo-doped

BiVO4 film for enhanced photoelectrochemical water oxidation. Chemical Engineering Journal. 2019;**378**: 122193

[146] Sharma P, Jang JW, Lee JS. Key Strategies to Advance the Photoelectrochemical Water Splitting Performance of α-Fe_2O_3 Photoanode. ChemCatChem. 2019;**11**:157

[147] Jiang C, Moniz SJA, Wang A, Zhang T, Tang J. Photoelectrochemical devices for solar water splitting - materials and challenges. Chemical Society Reviews. 2017;**46**:4645

[148] Baig F, Khattak YH, Soucase BM, Beg S, Ullah S. Effect of anionic bath temperature on morphology and photo electrochemical properties of Cu2O deposited by SILAR,. Materials Science in Semiconductor Processing. 2018;**88**:35

[149] Wang Q, Sun C, Liu Z, Tan X, Zheng S, Zhang H, et al. Ultrasound-assisted successive ionic layer adsorption and reaction synthesis of Cu_2O cubes sensitized TiO2 nanotube arrays for the enhanced photoelectrochemical performance. Materials Research Bulletin. 2019;**111**:277

[150] Desai MA, Sharma V, Prasad M, Jadkar S, Saratale GD, Sartale SD. Seed-layer-free deposition of well-oriented ZnO nanorods thin films by SILAR and their photoelectrochemical studies. International Journal of Hydrogen Energy. 2020;**45**:5783

[151] Jimenez-Gonzalez A, Suarez-Parra R. Effect of heat treatment on the properties of ZnO thin films prepared by successive ion layer adsorption and reaction (SILAR). Journal of Crystal Growth. 1996;**167**:649

[152] Patil UM, Gurav KV, Joo O-S, Lokhande CD. Synthesis of photosensitive nanograined TiO2 thin films by SILAR method. Journal of Alloys and Compounds. 2009;**478**:711

[153] Pathan HM, Min S-K, Desai JD, Jung K-D, Joo O-S. Preparation and characterization of titanium dioxide thin films by SILAR method. Materials Chemistry and Physics. 2006;**97**:5

[154] Shameem A, Devendran P, Siva V, Raja M, Asath Bahadur S, Manikandan A. Asath Bahadur, A. Manikandan, Preparation and Characterization Studies of Nanostructured CdO Thin Films by SILAR Method for Photocatalytic Applications. Journal of Inorganic and Organometallic Polymers. 2017;**27**:692

[155] Desai MA, Sartale SD. Facile Soft Solution Route to Engineer Hierarchical Morphologies of ZnO Nanostructures. Crystal Growth & Design. 2015;**15**:4813

[156] Senthamilselvi V, Saravanakumar K, Jabena Begum N, Anandhi R, Ravichandran AT, Sakthivel B, et al. Photovoltaic properties of nanocrystalline CdS films deposited by SILAR and CBD techniques-a comparative study. Journal of Materials Science: Materials in Electronics. 2012; **23**:302

[157] Shen T, Tian J, Lv L, Fei C, Wang Y, Pullerits T, et al. Investigation of the role of Mn dopant in CdS quantum dot sensitized solar cell, Electrochimica Acta. Electrochimica Acta. 2016;**191**:62-69

[158] Ravichandran K, Rajkumar P, Sakthivel B, Swaminathan K, Chinnappa L. Role of precursor material and annealing ambience on the physical properties of SILAR deposited ZnO films. Ceramics International. 2014;**40**: 12375

[159] Kumar KDA, Valanarasu S, Ganesh V, Shkir M, Kathalingam A, Alfaify S.

Effect of Precursors on Key Opto-electrical Properties of Successive Ion Layer Adsorption and Reaction-Prepared Al:ZnO Thin Films. Journal of Electronic Materials. 2018;**47**:2

[160] Shaikh AA, Waikar MR, Sonkawade RG. Effect of different precursors on electrochemical properties of manganese oxide thin films prepared by SILAR method, Synthetic Metals. Synthetic Metals. 2019;**247**:1-9

[161] Kumar P, Gowrish KR. The effect of precursor concentration and post-deposition annealing on the optical and micro-structural properties of SILAR deposited SnO_2 films. Materials Research Express. 2020;**7**:016428

[162] Hernández RG, Cruz MRA, Aguilar NP, Aguilar MER, López MQ, Guerra EM, Tostado FSA. A. Méndez-Vilas Ed. 431

[163] Kakade BN, Nikam CP, Gosavi SR. Effect of immersion cycles on structural, morphology and optoelectronic properties of nanocrystalline Ag2S thin films deposited by SILAR technique. IOSR Applied Physics. 2014;**6**(6):06-12

[164] Nwanya AC, Cosmas Chigbo SC, Ezugwu RU, Osuji M, Malik FI, Ezema A. Transformation of cadmium hydroxide to cadmium oxide thin films synthesized by SILAR deposition process: Role of varying deposition cycles. Arab Universities Basic and Applied Sciences. 2016;**20**:49

[165] Ashith VK, Gowrish RK. Structural and Optical Properties of ZnS Thin Films by SILAR Technique obtained by acetate Precursor. IOP Conference Series: Materials Science and Engineering. 2018; **360**:012058

[166] Preetha KC, Murali KV, Ragina AJ, Deepa K, Remadevi TL. Effect of cationic precursor pH on optical and transport properties of SILAR deposited nano crystalline PbS thin films. Current Applied Physics. 2012;**12**:53

[167] Sakthivelu A, Valanarasu S, Prince JJ. Effect of pH on SILAR deposited ZnO thin films. International Journal of Chemical Sciences. 2009;7(4):2463

[168] Patwary MAM, Hossain MA, Ghos BC, Chakrabarty J, Haque SR, Rupa SA, et al. Copper oxide nanostructured thin films processed by SILAR for optoelectronic applications. RSC Advances. 2022;**12**:32853

[169] Farhad SFU, Hossain MA, Tanvir NI, Akter R, Patwary MAM, Shahjahan M, et al. Structural, optical, electrical and photoelectrochemical properties of cuprous oxide thin films by modified SILAR method. Materials Science in Semiconductor Processing. 2019;**95**:68

[170] Farhad SFU, Majumder S, Hossain MA, Tanvir NI, Akter R, Patwary MAM. Patwary, Effect of Solution pH and post-annealing temperature on the properties of the Optical Bandgap of the Copper Oxide Thin Films Grown by modified SILAR Method. MRS Advances. 2019;**4** (16):937

[171] Gençyılmaz O, Taşköprü T. Effect of pH on the synthesis of CuO films by SILAR method. Journal of Alloys and Compounds. 2016;**695**:1205-1212

[172] Visalakshi S, Kannan R, Valanarasu S, Kathalingam A, Rajashabala S. Studies on optical and electrical properties of SILAR-deposited CuO thin films. Materials Research Innovations. 2016

[173] Cavusoglu H. Structural, morphological and optical studies of nanostructured cadmium oxide films: the role of pH. Journal of Materials

Science: Materials in Electronics. 2018; **29**:12777

[174] Lokhande CD, Patil PS, Tributsch H, Ennaoui A. Ennaoui, ZnSe thin films by chemical bath deposition method. Solar Energy Materials & Solar Cells. 1998;**55**:379

[175] Patwary MAM, Saito K, Guo Q, Tanaka T. Tanaka, Influence of oxygen flow rate and substrate positions on properties of Cu-oxide thin films fabricated by radio frequency magnetron sputtering using pure Cu target. Thin Solid Films. 2019;**675**:59-65

[176] Ashith VK, Rao GK, Moger SN, Smitha R. Effect of post-deposition annealing on the properties of ZnO films obtained by high temperature, micro-controller-based SILAR deposition. Ceramics International. 2018;**44**(15): 10669-10676

[177] Serin N, Serin T, Horzum S, Celik Y. Annealing effects on the properties of copper oxide thin films prepared by chemical deposition, Semicond. Semiconductor Science and Technology. 2005;**20**:398-401

[178] Tasdemirci TC. Copper Oxide Thin Films Synthesized by SILAR: Role of Varying Annealing Temperature. Elec tronic Materials Letters. 2020;**16**:239

[179] Gokul B, Matheswaran P, Sathyamoorthy R. Sathyamoorthy, Influence of Annealing on Physical Properties of CdO Thin Films Prepared by SILAR Method. Journal of Materials Science and Technology. 2013;**29**(1):17-21

[180] Iskenderoglu D, Guney H. Effect of Annealing on the Structural, Morphological and Optical Properties of MgO Nanowall Structures Grown by SILAR Method. Electronic Materials. 2019;**48**:5850-5856

[181] Patwary MAM, Saito K, Guo Q, Tanaka T, Yu KM, Walukiewicz W. Walukiewicz, Nitrogen Doping Effect in Cu4O3 Thin Films Fabricated by Radio Frequency Magnetron Sputtering. Physica Status Solidi B. 2020;**257**(2):1900363

[182] Patwary MAM, Ohishi M, Saito K, Guo Q, Yu KM, Tanaka T. Effect of Nitrogen Doping on Structural, Electrical, and Optical Properties of CuO Thin Films Synthesized by Radio Frequency Magnetron Sputtering for Photovoltaic Application. ECS Journal of Solid State Science and Technology. 2021;**10**:065019

[183] Ashcroft NW, Mermin ND. Solid State Physics Holt. New York: Rinehart and Winston; 1976

[184] Hossain MA, Patwary MAM, Rahman MM, Ohtsu Y. Properties of AZO thin films prepared by stationary and rotating RF magnetized plasma sputtering source. AIP Advances. 2022;**12**:015224

[185] Saha S, Jana M, Khanra P, Samanta P, Koo H, Murmu NC, et al. Band gap modified boron doped NiO/Fe3O4 nanostructure as the positive electrode for high energy asymmetric supercapacitors. RSC Advances. 2016;**6**:1380

[186] Zhu X, Zou X, Zhou H. Effects of Different Doping Ratio of Cu Doped CdS on QDSCs Performance. Journal of Nanomaterials. 2015;**498950**:4

[187] Belkhedkar MR, Ubale AU, Sakhare YS, Zubair N, Musaddique M. Musaddique, Characterization and antibacterial activity of nanocrystalline Mn doped Fe_2O_3 thin films grown by successive ionic layer adsorption and reaction method. Association of Arab Universities Basic & Applied Science. 2016;**21**:38

[188] Bayram O, Guney H, Ertargin ME, Igman E, Simsek O. Effect of doping

concentration on the structural and optical properties of nanostructured Cu-doped Mn3O4 films obtained by SILAR technique. Applied Physics A. 2018;**124**: 606

[189] F. Bayansal, T. Tas¸ Ko Pru, Bu Nyamin S, Ahin, Haci Ali C Etinkara. Effect of Cobalt Doping on Nanostructured CuO Thin Films. Metallurgical & Materials Transactions A, 45A, (2014) 3671

[190] Yuksel M, Pennings JR, Bayansal F, Yeow JTW. Effect of B-doping on the morphological, structural and optical properties of SILAR deposited CuO films. Physica B: Condensed Matter. 2020;**599**:412578

[191] Dhanabalan K, Ravichandran AT, Ravichandran K, Valanarasu S, Mantha S. Effect of Co doped material on the structural, optical and magnetic properties of Cu_2O thin films by SILAR technique. Materials Science: Materials in Electronics. 2017;**28**:4431

[192] Satheeskumar S, Vadivel S, Dhanabalan K, Vasuhi A, Ravichandran AT, Ravichandran K. Enhancing the structural, optical and magnetic properties of Cu_2O films deposited using a SILAR technique through Fe-doping. Journal of Materials Science: Materials Electronics. 2018;**29**:9354

[193] Iskenderoglu D, Güney H, Güldüren ME. Chromium - An effective dopant for engineering the structural and the optical properties of CdO nanostructures grown by SILAR method. Optical Materials. 2021;**115**:111067

[194] Sahin B, Bayansal F, Yüksel M. Influence of manganese concentration and annealing temperatures on the physical properties of CdO films grown by the SILAR method. Philosophical Magazine. 2014;**94**(9):956-963

[195] Yüksel M, Şahin B, Bayansal F. Nano structured CdO films grown by the SILAR method: Influence of silver-doping on the morphological, structural and optical properties. Ceramics International. 2016;**42**(5):6010-6014

[196] Ravichandran K, Banu NN, Baneto M, Selvi VS. Realizing near stoichiometric and highly transparent CdS: Mo thin films by a low-cost improved SILAR technique. Physica Status Solidi A: Applications and Materials Science. 2016;**213**(2):436-442

[197] Ravichandran K, Nisha Banu N, Senthamil Selvi V, Rajkumar PV. Realization of near stoichiometric Sn-doped CdS films through an improved SILAR Technique. Materials Research Innovations. 2015:1-6

[198] Gülen Y. Characteristics of Ba-Doped PbS Thin Films Prepared by the SILAR Method. Acta Physica Polonica A. 2014;**126**(03):763

[199] Sundhar A. Development of ZnS thin film with Co, Cu and Ag doping using SILAR Method. Materials Today: Proceedings. 2022;**48**(2):377-381

[200] Rajkumar PV, Ravichandran K, Baneto M, Ravidhas C, Sakthivel B, Dineshbabu N. Enhancement of optical and electrical properties of SILAR deposited ZnO thin films through fluorine doping and vacuum annealing for photovoltaic applications. Materials Science in Semiconductor Processing. 2015;**35**:189

[201] K. Radhi Devi, G. Selvan, M. Karunakaran, K. Kasirajan, L. Bruno Chandrasekar, Mohd Shkir, S. AlFaify, SILAR-coated Mg-doped ZnO thin films for ammonia vapor sensing applications. Journal of Materials Science Materials in Electronics, 2020

Chapter 4

Depth Profiling of Multilayer Thin Films Using Ion Beam Techniques

Mandla Msimanga

Abstract

Functional properties of thin film structures depend a lot on the thickness and chemical composition of the layer stack. There are many analytical techniques available for the identification and quantification of chemical species of thin film depositions on substrates, down to a few monolayers thickness. For the majority of these techniques, extending the analysis to several tens of nanometres or more requires some form of surface sputtering to access deeper layers. While this has been done successfully, the analysis tends to become quite complex when samples analysed consist of multilayer films of different chemical composition. Ion beam analysis (IBA) techniques using projectile ions of energies in the MeV range have a demonstrated advantage in the study of multilayer thin films in that the analysis is possible without necessarily rupturing the film, up to over 500 nm deep in some cases, and without the use of standards. This chapter looks at theoretical principles, and some unique applications of two of the most widespread IBA techniques: Rutherford Backscattering Spectrometry (RBS) and Elastic Recoil Detection Analysis (ERDA), as applied to multilayer thin film analyses.

Keywords: thin film, multilayer, ion beam analysis, ERDA, RBS, depth profiling

1. Introduction

Structure-property studies of thin films underpin research and development of new functional materials from fundamental experimental investigations right up to device fabrication stage. This is, to a large extent, made possible by the availability of specialised analysis tools able to probe materials at the nano/micrometre levels. Examples of analytical tools found in typical materials research labs include Atomic Force Microscopy (AFM), Scanning Electron Microscopy (SEM) for surface morphology, X-ray Photoelectron Spectroscopy (XPS), Auger Electron Spectroscopy (AES) for elemental and chemical state information, X-ray Diffraction (XRD) for crystal structure determination, Raman Spectroscopy and Fourier Transform Infrared Spectroscopy (FTIR) for molecular identification, and so on [1, 2]. Thin film coatings of up to a few 100nm thickness abound in many advanced technological applications, including sensor devices designed for a whole range different stimulus [3]. These thin film structures derive their functional properties from their physical dimensions and chemical makeup. Film thickness, for instance, plays a key role in semiconductor solar

IntechOpen

radiation detectors, in determining the fraction of solar radiation that is absorbed in the active region of the detector [4]. The concentration and depth distribution of dopant species in semiconductor materials is key to the operation of sensor devices based on thin film diode and/or transistor structures. The aforementioned analytical techniques can readily provide surface and structural properties of a film but not so much thickness and elemental depth profile information. Ion beam analysis techniques using MeV energy beams have a demonstrated capability to provide this information, without the use of standards in most instances [5]. At the highest level of performance, standard-free analysis at 1% traceable accuracy has been reported [6].

Multilayer structures present unique challenges for elemental depth profiling analytical techniques. In sputter depth profiling using any of XPS, AES, or Secondary Ion Mass Spectrometry (SIMS) there is a need for standards for calibration of the sputter etch rate to a depth scale. On the other hand, the interaction depth/range of MeV ions allows for probing films to depths of up to 1 micrometre or more depending on the probing ion species and energy, without necessarily sputtering the material. For a film comprising different layers the ion-matter interaction parameters change in a way that makes it possible to distinguish the different layers, again without recourse to a reference standard. This chapter begins with a look at how these fundamental interactions are exploited in two widely used ion beam analysis techniques; Rutherford Backscattering Spectrometry (RBS) and Elastic Recoil Detection Analysis (ERDA). The discussion then progresses to description of a typical experimental set up before looking at practical multilayer film analysis examples that showcase the unique strengths of IBA techniques.

2. Ion-matter interactions at MeV energies

When a swift ion penetrates solid matter a number of interactions may take place between the ion and the target atomic nuclei and electrons. These include elastic and inelastic scattering, nuclear reactions, excitation and ionisation, photon emission, etc. The extent of the interaction between a beam of ions and the target atoms depends largely on the particular collision cross section σ (E, Z_1, Z_2), which in many instances is a function of both the ion energy and atomic numbers of the incident and target atoms. The cross-section gives the probability of a given type of ion-atom interaction taking place. If the cross section for a particular ion-atom combination is known, then detecting, counting and sorting the products of an interaction in some systematic way can provide information about the nature of target atoms. This information could be any of the identity, concentration and depth distribution of a particular atomic species in a film. Therein lies the core of ion beam analysis techniques, and indeed other similar 'probe-and-measure' analytical techniques. Within the kaleidoscope of possible interactions that may occur when an ion beam strikes a solid target material, there are three main physical parameters that underpin the application of ion beams in materials analysis. These are described hereunder.

2.1 Kinematics of ion-atom collisions

As a swift ion moves through solid matter, it interacts with both the electron cloud and the atomic nuclei of the target material. Collisions with target nuclei provide the basis for identification of the target atom species according to their mass. This is achieved through treating the interaction as a binary elastic collision between the two

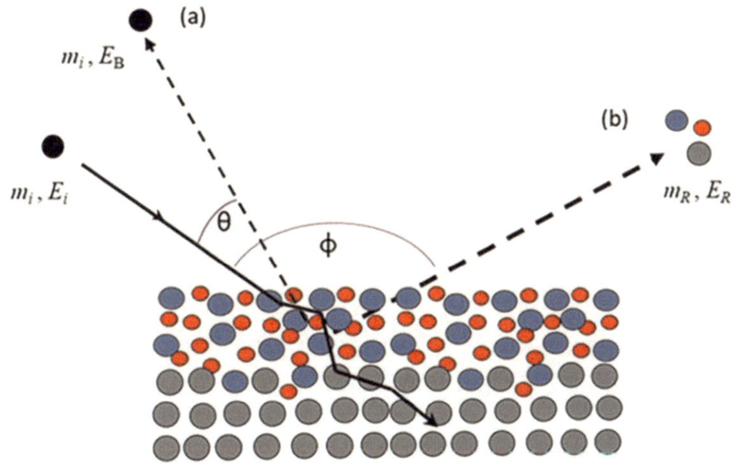

Figure 1.
Collision kinematics for (a) RBS and (b) ERDA techniques. In RBS an incident ion of mass m_i with energy E_i is backscattered through an angle $0 < \theta < 90$ with an energy E_B, and in ERDA target atoms of different masses m_R, are forward recoiled at an angle $90^o < \varphi < 180^o$ with different energies E_R. The actual values of θ and φ are defined by the orientation of the RBS and ERDA detectors.

particles. In the ideal case the effects of the electron sub-system and the neighbouring nuclei are ignored, thereby simplifying the mathematical treatment [7]. **Figure 1** shows a representation of the ion-atom collisions relevant to (a) RBS and (b) ERDA. The backscattering and recoil angles are defined with respect to the incident beam direction by the orientation of the detectors employed for the measurement.

For RBS, applying the principle of conservation of kinetic energy and momentum in elastic collisions leads to the following equation that gives the backscattering energy E_b, of the incident ion of mass m_i, after collision with a target atom of mass m_t:

$$\frac{E_b}{E_i} = k_b = \left(\frac{-m_i \cos\theta \pm \sqrt{m_t^2 - m_i^2 \sin^2\theta}}{m_i + m_t} \right)^2 \tag{1}$$

where k_b is known as the backscattering kinematic factor, with the plus sign in Eq. (1) holding for $m_i < m_t$. A similar consideration for **Figure 1b** leads to:

$$\frac{E_r}{E_i} = k_r = \frac{4m_i m_r}{(m_i + m_r)^2} \cos^2\phi \tag{2}$$

where, k_r is referred to as the recoil kinematic factor describing the ratio of the recoil atom energy E_r, to the incident ion energy E_i, and m_r is the mass of the recoil atom. The angles θ and φ are as defined in **Figure 1**. In principle then, for a given experimental configuration, measurement of the energy of the backscattered ion in RBS or that of the recoiled atom in ERDA can be used to determine the mass of the target or recoil atom, respectively.

2.2 Collision cross section

The quantitative capability of ion beam analysis techniques is a direct consequence of the physical concept of cross section(σ) in ion-target interactions. The cross section describes the probability of a backscattering or recoil event occurring in a given direction, defined by the detector solid angle (Ω), and is a function of the interaction potential associated with the collision. Continuing with the concept of point charge interactions used in kinematics, quantitation in both RBS and ERDA techniques is based on a Coulomb interaction potential. For a pure Coulomb potential the RBS or Rutherford scattering cross section is given by [7, 8].

$$\left(\frac{d\sigma}{d\Omega}\right)_{scattered} = \left(\frac{Z_i Z_t e^2}{2E_i}\right)^2 \frac{\left(\sqrt{m_t^2 - m_i^2 \sin^2\theta} - m_2 \cos\theta\right)^2}{m_t \sin^4\theta \sqrt{m_t^2 - m_i^2 \sin^2\theta}} \tag{3}$$

similarly the recoil cross section is given by

$$\left(\frac{d\sigma}{d\Omega}\right)_{recoil} = \left(\frac{Z_i Z_r e^2}{2E_i}\right)^2 \left(1 + \frac{m_i}{m_r}\right)^2 \frac{1}{(-\cos^3\phi)} \tag{4}$$

where Z_i, Z_t and Z_r are the atomic numbers of the incident, target and recoil ions, respectively, and e is the electron charge. The angles θ and φ are again as defined in **Figure 1**. The concentration of a given atomic species in a sample is then obtained from the experimental yield, which is directly proportional to the cross section, in the energy spectrum associated with that element. Real collisions approximate this simplified approach in cases where the incident particle is totally stripped of its electrons and can thus be regarded as a point charge. Corrections are nonetheless needed to account for deviations from the ideal case scenario in instances where the incident ion energy is low, to a point that screening by orbital electrons cannot be neglected. Anderson and co-workers [9] reviewed the energies at which this deviation occurs and these are now fairly well known. Deviation also occurs at the high-energy side when the distance of closest approach of the nuclei is within the range of nuclear forces and the interaction potential is no longer a simple Coulomb potential. Bozoian et al. [10] have determined the energies at which nuclear force field effects become significant.

Additional parameters that are needed for concentration evaluation are the incident beam dose and the detector solid angle. Eqs. (3) and (4) both show that for a given experimental geometry, the experimental yield is directly proportional to the square of the Z-values and inversely proportional to the square of the incident beam energy. This fact underpins the preference of low energy heavy ions in the case of ERDA, as this favours good measurement statistics within a relatively short measurement time. For RBS, while the cross section is higher for heavier and slower incident ions, the requirement that $m_i < m_t$ precludes their use in many applications. This limitation is generally countered through delivering fairly high beam doses of light element projectiles to get acceptable spectral yields.

2.3 Energy loss rate: Stopping force

Depth analysis in ion beam analytical techniques follows directly from the energy lost by the projectile and the target atoms traversing the target sample [11]. In RBS for example, the energy loss of the incident ion as it enters and exits the target sample gives the location of the scattering atom below the surface, whereas in ERDA it is the total energy lost by the projectile ion as it enters and the recoil atom as it exits the target sample that gives a similar indication. The energy loss per unit depth, or the stopping force, is a fundamental ion-atom interaction parameter that is the key linkage between an energy spectrum and the thickness of a target layer. In basic terms the energy width dE in a measured energy spectrum depends on the thickness Δx of a target layer according to:

$$\Delta x = \int \frac{1}{S(E)} dE \quad (5)$$

where S(E) is the energy dependent stopping force. The energy loss of the incident ion traversing a target material arises from two types of interactions which are dependent on the ion velocity [12]. At low energies (below about 1 keV/u) the energy loss is mainly due to elastic collisions with target nuclei. This is referred to as nuclear energy loss. As energy increases, interaction with the electron cloud becomes more dominant as the ion speed approaches that of the orbital electrons in target atoms. The energy loss is then mainly through inelastic collisions with the target electrons of the system. This is referred to as electronic energy loss and for the typical ion energies used in IBA, this is the dominant mode of energy loss.

Stopping force is also dependent on the Z-values of the colliding particles. For a fixed target, the stopping force increases with the projectile ion charge at the same velocity. This points to better depth resolution when heavy ions are used, according to Eq. (5). This, however, is at the expense of ion range or analytical depth since heavy ions would have a shallower range because of the higher stopping force. There are several theoretical formulations that are of semi-empirical [13] or *ab initio* [14, 15] origin that are used to calculate S(E) for a wide range of ion-atom combinations and energy ranges. It goes without saying that the accuracy of the stopping force data that is used in the energy-to-depth calculations is one of the major contributing factors to the accuracy of layer thickness and depth profile measurements.

3. Analytical software

The three physical concepts of binary collisions, collision cross section and stopping force discussed above constitute the theoretical foundation of ion beam analysis techniques. Practical implementation of the techniques requires taking into account many other additional effects that lead to deviations from the ideal situation. Software codes have been devised over the years to aid the interpretation of energy spectra obtained in a measurement. These can be broadly classified into two categories. The first category encompasses codes that calculate concentration profiles directly from experimental energy spectra using modified analytical calculations [16, 17]. These modifications could include for instance screened Rutherford cross sections, where the effect of the electron cloud on the pure Coulomb interaction is taken into account in ion-atom collisions. One advantage of these codes is that they are unlikely to generate more information about the sample than the data actually contain. The main drawback is that the profiles derived are, strictly speaking, not real concentration depth profiles since they still have the energy resolution convoluted with effects of the actual sample structure and so one has to live with the limited depth resolution given by the experimental and sample conditions.

The second category (mainstream codes are in this group) includes codes that tackle the problem from the opposite viewpoint; a hypothetical sample structure is assumed and a theoretical energy spectrum calculated either analytically, as in SIMNRA [8], or by Monte Carlo (MC) methods [17, 18] and compared with the experimental one. This sample structure is altered until a best fit is obtained between the simulated and the experimental- spectra [18]. This iterative simulation approach uses analytical functions to convolute the ideal energy spectrum so as to consider most of the physical limitations that include the detector resolution, energy loss straggling, multiple scattering and sample roughness [11] among others.

Monte Carlo simulation-based codes such as MCERD, that is in-built in *Potku* [17] and CORTEO [18] stand apart from the deterministic codes in the sense that MC methods principally include all the important phenomena involved in ion-target interactions. The approach employed here is that the calculation follows individual ion trajectories to negligible energies, based on analytical functions that describe ion stopping and collision cross sections with the necessary correction (e,g. screening functions) implemented. In this way complex physical processes such as multiple scattering and the interaction between ions and the detector system are taken into account in a natural way, without the approximations that analytical codes involve. The one drawback of analytical and Monte Carlo simulation codes is that they may include phantom structural details that lead to a good fit to experimental data but not necessarily reflecting the true sample structure. Good practice in IBA then dictates using both direct calculation and iterative simulation codes to get a more accurate interpretation of the measurement data.

4. IBA instrumentation

Ion beam techniques are based on high energy ion beams generated from particle accelerators. Typical IBA accelerators vary in size, from small compact tandetrons of 2.0 MV terminal voltage to fairly huge tandem accelerators of up to 20 MV [19]. These machines deliver particles of energies ranging from 0.1–to 10 MeV/u, depending on particle mass. In brief, specific ions are injected from the ion source into the accelerator column where particle acceleration is due to a huge electrostatic field. On exiting the acceleration stage, a magnetic field is used to select ions of a specific charge to mass ratio, or energy, to filter through to the experimental end station or scattering chamber. The schematic in **Figure 2** shows the general set up for both RBS and ERDA analysis techniques. There are of course additional accessory systems such as beam diagnostics, beam focusing elements, vacuum systems and high voltage power supplies, and so on that make up a complete accelerator system.

As pointed out in the introduction, functional exploitation of ion-matter interactions in ion beam analysis depends on the positioning and type of particle and/or photon detectors to detect and count the relevant products of ion-atom collisions. In RBS for instance, solid state semiconductor detectors are generally used to count the number of incident ions backscattered through a particular angle and to measure their energy as well. Raw data is collected in the form of an energy spectrum of backscattered particles. It is this energy spectrum that is fitted using analytical software like SIMNNRA [8] to extract sample properties such as elemental depth profiles.

In the case of ERDA, two detector variants are available. The simplest or conventional set up, mostly used for hydrogen analysis, consists of a solid-state detector with a filter foil in front of it to stop all other atomic species besides hydrogen. It is possible though, to select the incident beam species, energy and filter foil in such a way that other ions heavier than hydrogen can be analysed—if hydrogen itself is not one of the

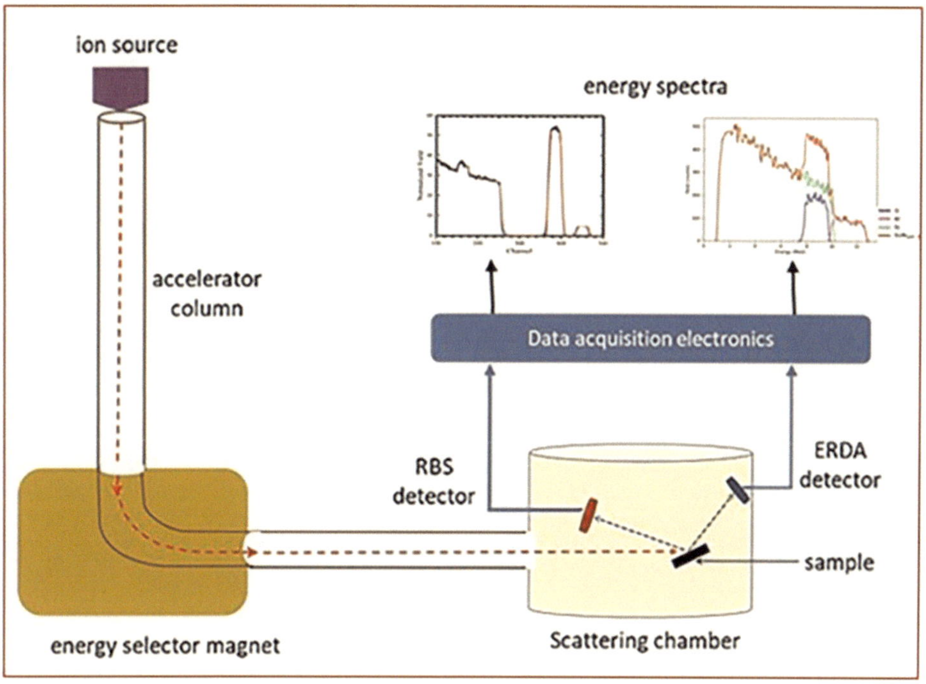

Figure 2.
Basic set up for RBS and ERDA ion beam analysis techniques.

constituent elements of the target sample. The limitation of the conventional set up is quite apparent. If the object of analysis is the depth distribution of several elements in a sample then this configuration cannot be used. Mass dispersive detector systems such as time-of-flight (ToF) telescopes [20] become quite useful in this regard. In a ToF detector set up the energy of recoil atoms is measured simultaneously with their

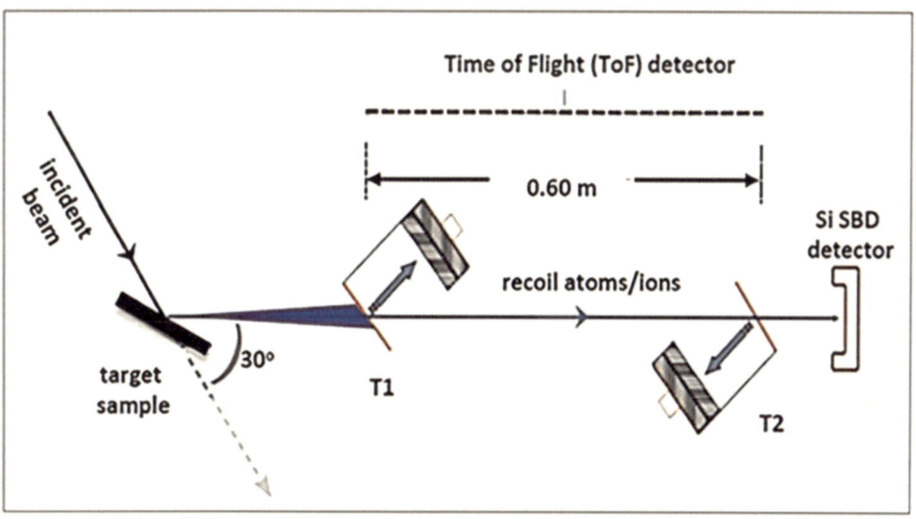

Figure 3.
The time-of-flight detector set up for heavy ion ERDA at iThemba LABS.

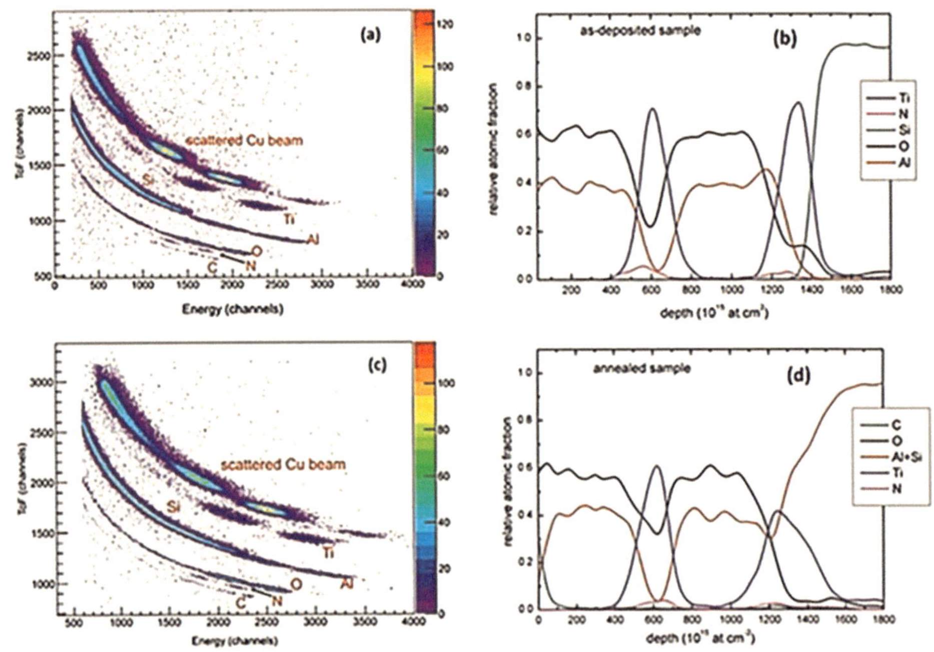

Figure 4.
ToF vs. energy scatter plots from the analysis of an Al_2O_3-Ti bi-layer stack on a silicon substrate before annealing (a) and after annealing (c) in vacuum at 800°C. the resultant depth profiles are shown in (b) and (d), respectively. Channels in (a) and (b) refer to as yet uncalibrated time and energy axes. (reproduced with permission from ref. [20]).

transit time over a known distance—leading to mass identification or separation. **Figure 3** is a schematic of the ToF detector system used at the iThemba LABS ERDA set-up, where the flight path is 0.6 m long. Raw data generated from such a detector set up consists of 2-D scatter plots of ToF vs. energy (**Figure 4**) from which elemental energy spectra can be extracted and fed into analytical software for either direct depth profile calculation [16, 17] or simulation [8, 21].

5. Application examples

The examples of analyses below have been selected from the literature to highlight some of the relative strengths (and weaknesses) of the two IBA techniques discussed in this chapter.

5.1 ERDA analysis of an alumina-titanium layer stack

Figure 4 shows results of ERDA depth profiling of an Al_2O_3/Ti/Al_2O_3/Ti/Si bi-layer film annealed at 800°C in vacuum [20], to study the Al_2O_3-Ti solid state reaction. The measurement was carried out using a 26.1 MeV $^{63}Cu^{7+}$ beam, with a time-of-flight (ToF) recoil detector mounted at 30° (φ = 150° in **Figure 1**) to the incident beam direction. Further details of the measurement set-up are given in Msimanga et al. [20]. Coincidence measurement of the ToF and energy of atoms recoiled from the target sample allows for their separation according to mass. The 2-D scatter plots in

Figures 4a and **c** show all the elements detected from each target sample, as well as the forward scattered incident ^{63}Cu beam. The shape of the scatter plots derives from the simple inverse relationship between the ToF and kinetic energy E_r of the recoil atoms; $ToF = \sqrt{m_r/2\,E_r}{\bullet}L$, where m_r is the recoil atom mass and L is the length of the flight path. For a given atomic species, atoms recoiled right from the surface will be detected with higher energy (i.e. shorter ToF) than those from deeper layers due to energy loss of the latter as they move through, and out of the sample. And for a given energy E_r, the ToF increases as the particle mass increases, i.e. heavier atoms move slower, hence the observed separation of recoil atoms in terms of mass.

The depth profile of the as-prepared sample is shown in **Figure 4b** and that of the annealed one in **Figure 4d**. To get a sense of the analytical depth shown, if the depth scale is converted to units of nanometres using known atomic densities of the Al_2O_3 and Ti, the Ti/Si interface is about 150 nm from the surface in **Figure 4b** and shifts to just under 140 nm in **Figure 4d**, indicating silicide formation at the interface. The advantage of such 'raw' depth profiles, calculated using the direct energy-to-depth conversion code KONZERD [16] is that they give a quick visual description of the layer structure, with no need for a standard or reference sample. This provides a good starting point for further, more detailed analysis through MC simulation codes as described in ref. [20].

5.2 RBS analysis of $Ti_{0.7}Al_{0.3}N$/MoN and CrN/MoN multilayer films

Bin Han et al. [22] report of the analysis of $Ti_{0.7}Al_{0.3}N$/MoN and CrN/MoN multilayer films on Si(001) substrates using RBS. The incident ions used were 2.42 MeV and 1.52 MeV Li^{2+} ions at varying incident angles. The RBS data was supplemented by XPS, SEM and HR-TEM. The measurement using the 2.42 MeV energy beam at 0° incidence was able to resolve the Ti and Mo signals from first seven pairs of bi-layers of 44 nm $Ti_{0.7}Al_{0.3}N$ and 32 nm MoN. **Figure 5** shows the experimental and SIMNRA simulated spectra for extracting the depth profiles [22]. For Al this was possible only for the first three bilayers—beyond which the overlap of the Al signal with that of Ti and Mo from deeper layers precluded Al analysis. A pertinent observation that Bin Han and co-workers make from their results is that the low energy incident beam gave a much a higher backscattering yield (an indication of enhanced sensitivity—see Eq. (3)) and better depth resolution, but for a shallower analytical depth. For a fixed beam energy, increasing the tilt angle improved the depth resolution but again at the expense of the analytical depth. Another important observation that highlights a chink in the armoury of RBS is that while the surface sensitive XPS confirmed nitrogen in the topmost layers, it also pointed out presence of oxygen in those layers. RBS could not, because of the 'shadowing' effect of the (heavier) substrate element signal on that of light elements.

5.3 RBS and ERDA analysis of a solar thermal absorber stack

In *A study of solar thermal absorber stack based on* $CrAlSiN_x/CrAlSiN_xO_y$ *structure by ion beams*, AL-Rjoub et al. [23] describe RBS and ERDA measurements of a four-layer solar absorber film stack. The RBS data leads to rather inconclusive findings due to extensive overlap of signals from the different layers. On the other hand, the mass dispersive detector of the ToF-ERDA system allows separation of all the elements in the layer stack according to mass. Depth profiles are then extracted from Monte Carlo simulation using the MCERD code [17].

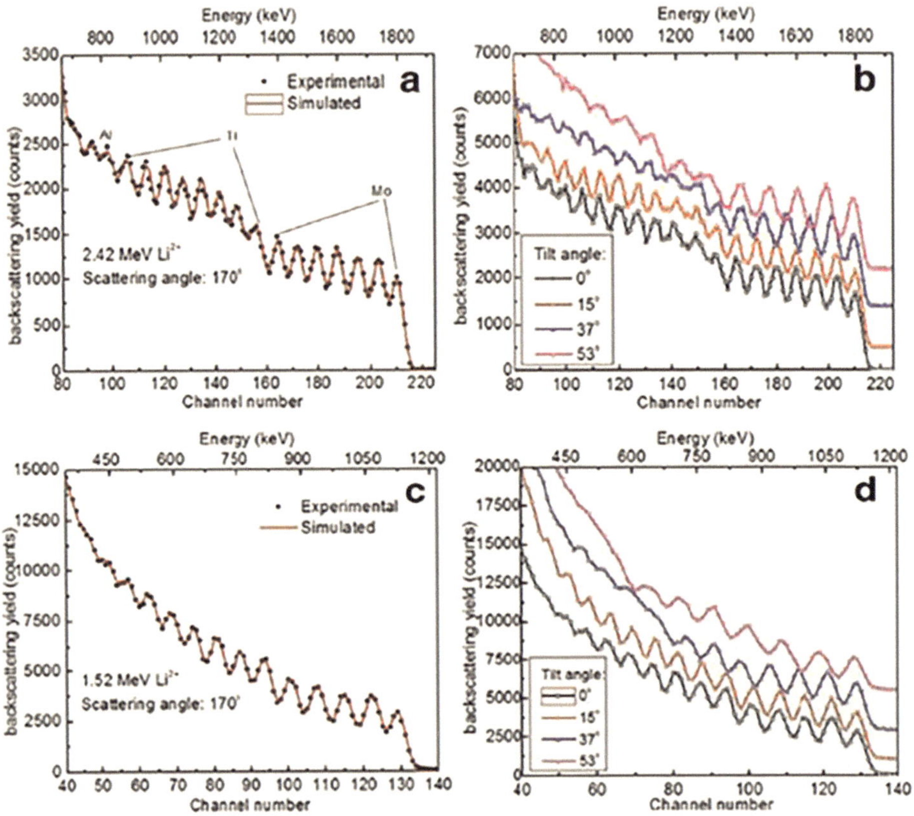

Figure 5.
RBS spectra from the analysis of $Ti_{0.7}Al_{0.3}N$/MoN and CrN/MoN multilayer films using 2.42 MeV Li2+ ions at 0° incidence angle (a) and at varying tilt angles (b). Similar spectra are shown in (c) and (d) respectively for a 1.52 MeV Li^{2+} incident beam. (taken from ref. [22], reproduced under the terms of the creative commons CC BY licence).

Figure 6 shows, raw data from ToF and energy detectors, experimental and simulated energy spectra of oxygen recoils from different layers and, the depth distribution of oxygen in the layer stack. While the obtained profiles are rather unrealistic step functions, it is widely accepted that due to the *one ion at a time* nature of the treatment of ion-target collisions in MC simulations, they produce the closest description to reality of a target structure.

5.4 Real-time RBS analysis of a hydrogen storage Pd/Ti/Pd layer stack

Another unique application of IBA techniques is the study of solid state reactions in real-time. This is a technique which has been pioneered by among other labs, the iThemba LABS (formerly known as the National Accelerator Centre) in South Africa [24]. Magogodi et al. [25] report of in-situ real-time RBS analysis of a 125 nm thick Pd/Ti/Pd film stack to investigate diffusion kinetics and stoichiometric evolution under different annealing environments. This formed part of a study aimed at developing hydrogen storage materials. The measurement described used 2 MeV He^{2+} ions to probe the layer structure as the samples were annealed in vacuum and in hydrogen environments, with the data taking starting from 160°C up to 600°C, at 30-second intervals.

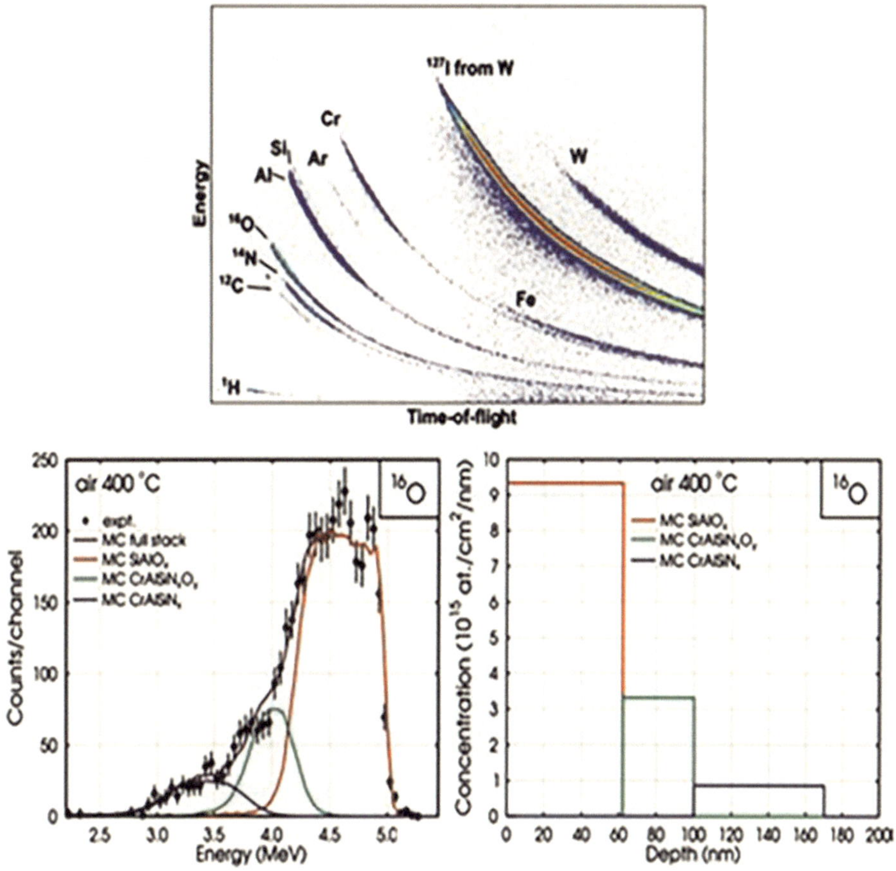

Figure 6.
Raw energy vs. ToF data (left), experimental and simulated oxygen energy spectra (Centre), and depth distribution of oxygen (right) in a $SiAlO_x/CrAlSiN_xO_y/CrAlSiN_x/W$ layer stack. (reproduced with permission from ref. [23]).

Figure 7 is a 2-D plot of the colour coded spectral yield as the temperature increases. For the sample annealed in vacuum, **Figure 7a**, the authors surmise that there is complete reaction between the Pd and Ti layers by the time the sample temperature reaches 550°C, and for the one annealed in a H_2 environment, their conclusion is that the Ti-Pd reaction is, to a great extent, inhibited. The obvious advantage of such a measurement is that by monitoring the reaction in real-time, any intermediate phases that may form are also detected, and not just the end-point—thus facilitating the study of solid-state reaction mechanisms. Indeed this has been applied in, for example, studies of growth kinetics of Ni(Pt) silicides [26], where the analysis of the huge data generated was done using artificial neural networks.

6. Summary

Multilayer thin film structures have become ubiquitous in many device structures in the current era of nanotechnology driven advances in electronics, medicine, energy

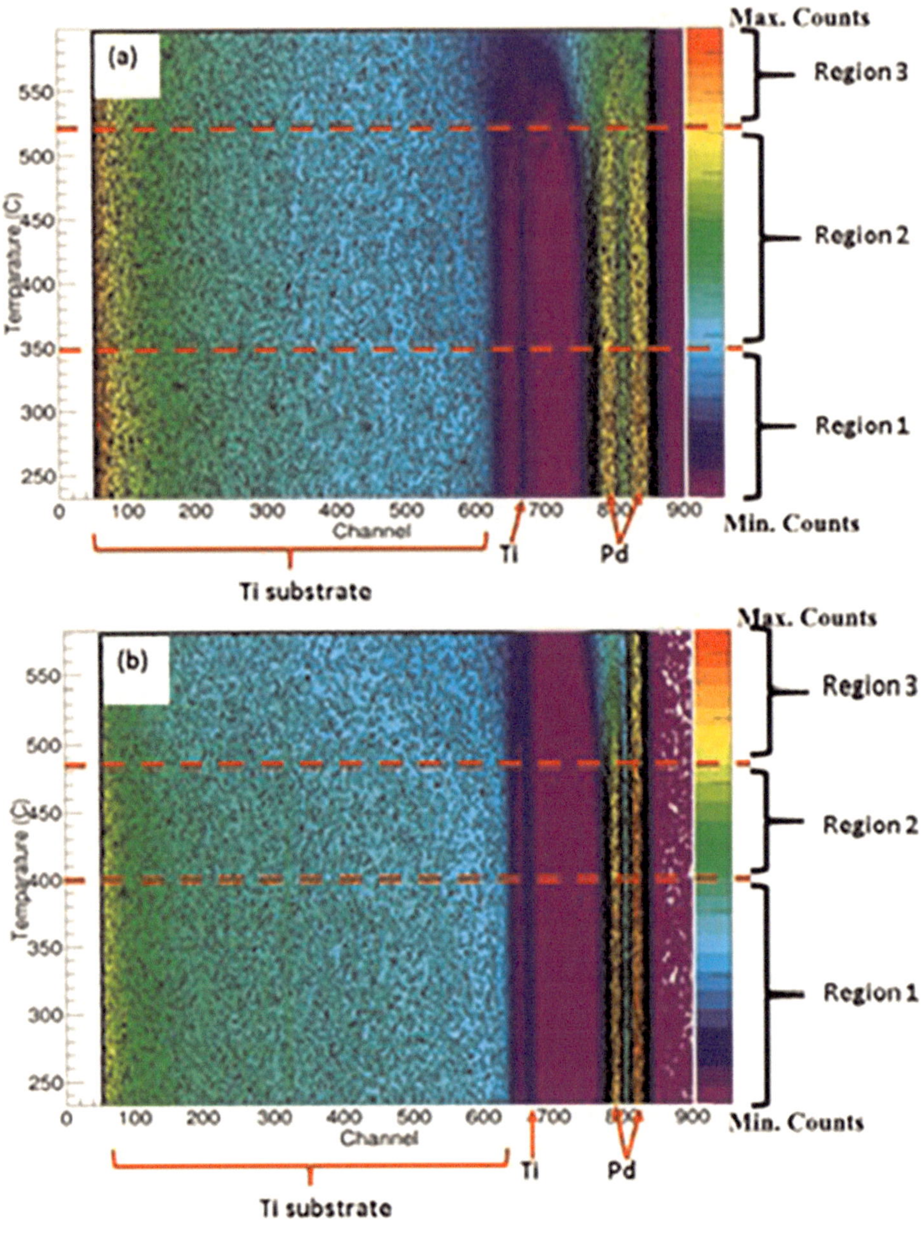

Figure 7.
Contour plots comparing the onset of reaction between atomic species in Pd/Ti/Pd layers annealed in (a) vacuum and in (b) hydrogen environments. (reproduced with permission from ref. [25]).

and other technological fields. Ion beam analysis techniques can readily generate invaluable structural information about multilayer films through standard-free analyses that are not possible with other analytical techniques. The physics behind RBS and ERDA techniques described in this chapter is well established and there are data analysis tools available to the materials analyst that can facilitate interpretation of experimental data with reasonably good accuracy. The selected applications described showcase the versatility of these analytical tools in addressing different problems from simple film thickness measurements to tracking solid state reactions in real-time.

Acknowledgements

The author would like to gratefully acknowledge financial and infrastructural support from the Tshwane University of Technology through the Photovoltaic Nanocomposites Research and Development Platform, and from NRF-iThemba LABS.

Conflict of interest

The author declares no conflict of interest.

Author details

Mandla Msimanga[1,2]

1 Department of Physics, Tshwane University of Technology, Pretoria, South Africa

2 NRF-iThemba LABS TAMS, Johannesburg, South Africa

*Address all correspondence to: msimangam@tut.ac.za

References

[1] Vickerman JC, Gilmore IS, editors. Surface Analysis: The Principal Techniques. 2nd ed. Chichester, UK: Wiley; 2011. p. 688

[2] O'Connor DJ, Sexton BA, Smart R, St C. Surface Analysis Methods in Materials Science. 2nd ed. Berlin, Heidelberg: Springer; 2003. p. 585. DOI: 10.1007/978-3-662-05227-3

[3] Tiginyanu I, Topala P, Ursaki V. Nanostructures and Thin Films for Multifunctional Applications: Technology, Properties and Devices. Switzerland: Springer; 2016. p. 576. DOI: 10.1007/978-3-319-30198-3

[4] Zhang D, Fan B, Ying L, Li N, Brabec CJ, Huang F, et al. Recent progress in thick-film organic photovoltaic devices: Materials, devices, and processing. Susmat WILEY. 2021;**1**: 4-23. DOI: 10.1002/sus2.10

[5] Jeynes C, Colaux JL. Thin film depth profiling by ion beam analysis. Analyst. 2016;**141**:5944. DOI: 10.1039/c6an01167e

[6] Jeynes C, Barradas NP, Szilagyi E. Accurate determination of quantity of material in thin films by Rutherford backscattering spectrometry. Analytical Chemistry. 2012;**84**:6061-6069. DOI: 10.1021/ac300904c

[7] Jeynes C, Webb RP, Lohstroh A. Ion beam analysis: A century of exploiting the electronic and nuclear structure of the atom for materials characterization. Reviews of Accelerator Science and Technology. 2011;**4**:41-82. DOI: 10.1142/S1793626811000483

[8] Mayer M. SIMNRA User's Guide. Report IPP 9/113, Max-Planck-Institut fur Plasmaphysik. Germany: Garching; 1997

[9] Andersen HH, Besenbacher F, Loftager P, Moller W. Large angle scattering of light ions in the weakly screened Rutherford region. Physical Review A. 1891;**21**:1980

[10] Bozoian M, Hubbard KM, Nastasi M. Deviations from Rutherford-scattering cross sections. Nuclear Instruments and Methods in Physics Research Section B. 1990;**51**:311-319

[11] Szilagyi E. On the limitations introduced by energy spread in elastic recoil detection analysis. Nuclear Instruments and Methods in Physics Research Section B: Beam Interactions with Materials and Atoms. 2001;**183**: 25-33

[12] Sigmund P. Particle Penetration and Radiation Effects. General Aspects and Stopping of Swift Point Particles. Springer Series in Solid State Sciences. Berlin: Springer-Verlag; 2006. p. 439. DOI: 10.1007/3-540-31718-X

[13] Ziegler JF, Ziegler MD, Biersack JP. SRIM—The stopping and range of ions in matter. Nuclear Instruments and Methods in Physics Research Section B. 2010;**268**:1818-1823. DOI: 10.1016/j.nimb.2010.02.091

[14] Schinner A, Sigmund P. Expanded PASS stopping code. Nuclear Instruments and Methods in Physics Research Section B: Beam Interactions with Materials and Atoms. 2019;**460**: 19-26. DOI: 10.1016/j.nimb.2018.10.047

[15] Grande PL, Schiwietz G. Convolution approximation for the energy loss, ionization probability and straggling of fast ions. Nuclear Instruments and Methods in Physics Research Section B: Beam Interactions

with Materials and Atoms. 2009;**267**: 859-863. DOI: 10.1016/j.nimb.2009.02.017

[16] Bergmaier A, Dollinger G, Frey CM, Faestermann T. Quantitative elastic recoil detection (ERD). Fresenius' Journal of Analytical Chemistry. 1995; **353**:582-584. DOI: 10.1007/BF00321328

[17] Arstila K et al. Potku, new analysis software for heavy ion elastic recoil detection analysis. Nuclear Instruments and Methods in Physics Research Section B: Beam Interactions with Materials and Atoms. 2014;**331**:34-41. DOI: 10.1016/j.nimb.2014.02.016

[18] Barradas NP et al. International atomic energy agency intercomparison of ion beam analysis software. Nuclear Instruments and Methods in Physics Research Section B: Beam Interactions with Materials and Atoms. 2007;**262**: 281-303. DOI: 10.1016/j.nimb.2007.05.018

[19] IAEA Interactive Map of Accelerators [Internet]. 2022. Available from: https://nucleus.iaea.org/sites/accelerators/Pages/Interactive-Map-of-Accelerators.aspx

[20] Msimanga M, Wamwangi D, Comrie CM, Pineda-Vargas CA, Nkosi M, Hlatshwayo T. The new heavy ion ERDA set up at iThemba LABS Gauteng: Multilayer thin film depth profiling using direct calculation and Monte Carlo simulation codes. Nuclear Instruments and Methods in Physics Research Section B: Beam Interactions with Materials and Atoms. 2013;**296**: 54-60. DOI: 10.1016/j.nimb.2012.11.015

[21] Schiettekatte F. Fast monte carlofor ion beam analysis simulations. Nuclear Instruments and Methods in Physics Research Section B: Beam Interactions with Materials and Atoms. 2008;**266**: 1880-1885. DOI: 10.1016/j.nimb.2007.11.075

[22] Han B et al. RBS depth profiling analysis of (Ti, Al)N/MoN and CrN/MoN multilayers. Nanoscale Research Letters. 2017;**12**:161. DOI: 10.1186/s11671-017-1921-3

[23] AL-Rjoub A, Costa P, Rebouta L, Bogdanovic Radovic I, Arstila K, Barradas NP, et al. A study of solar thermal absorber stack based on $CrAlSiN_x/CrAlSiN_xO_y$ structure by ion beams. Nuclear Instruments and Methods in Physics Research Section B: Beam Interactions with Materials and Atoms. 2019;**450**:195-199. DOI: 10.1016/j.nimb.2018.04.024

[24] Theron CC, Mars JA, Churms CL, Farmer J, Pretorius R. In situ, real-time RBS measurements of solid state reaction in thin films. Nuclear Instruments and Methods in Physics Research Section B: Beam Interactions with Materials and Atoms. 1998;**139**:213-218. DOI: 10.1016/S0168-583X(97)00946-4

[25] Magogodi S et al. Effect of annealing atmosphere on the layer interdiffusion in Pd/Ti/Pd multilayer stacks deposited on pure Ti and Ti-alloy substrates. Nuclear Instruments and Methods in Physics Research Section B: Beam Interactions with Materials and Atoms. 2019;**461**: 37-43. DOI: 10.1016/j.nimb.2019.09.018

[26] Demeulemeester J, Smeets D, Comrie CM, Barradas NP, Vieira A. On the growth kinetics of Ni(Pt) silicide thin films. Journal of Applied Physics. 2013;**113**:163504. DOI: 10.1063/1.4802738

Section 2

Applications of Thin Films

Chapter 5

Encapsulation of Perovskite Solar Cells with Thin Barrier Films

Katherine Lochhead, Eric Johlin and Dongfang Yang

Abstract

Long-term stability is a requisite for the widespread adoption and commercialization of perovskite solar cells (PSCs). Encapsulation constitutes one of the most promising ways to extend devices for lifetime without noticeably sacrificing the high power conversion efficiencies that make this technology attractive. Among encapsulation strategies, the most investigated methods are as follows: (1) glass-to-glass encapsulation, (2) polymer encapsulation, and (3) inorganic thin film encapsulation (TFE). In particular, the use of UV-, heat-, water-, and/or oxygen-resistant thin films to encapsulate PSCs is a new and promising strategy for extending devices for lifetime. Thin films can be deposited directly onto the PSC, as in TFE, or can be used in conjunction with glass-to-glass and polymer encapsulation to effectively prevent the photo-, thermal-, oxygen-, and moisture-induced degradation of the perovskite. This chapter will outline perovskite degradation mechanisms and provide a summary of the progress made to-date in the encapsulation of PSCs, with a particular focus on the most recent and promising advances that employ thin films. Additionally, the strengths and limitations of TFE approaches will be identified and contrasted against existing encapsulation strategies. Finally, possible directions for future research that can further enhance encapsulation effectiveness and extend PSC for lifetimes towards the 25-year target will be proposed.

Keywords: perovskite solar cells, thin film encapsulation, perovskite degradation, thin films, photovoltaics

1. Introduction

In the history of all photovoltaic technologies, the swift evolution of perovskite solar cells (PSCs) remains completely unprecedented. With the achievement of efficiencies that have increased from 14% to as high as 25.7% in less than 10 years [1], PSCs are on the verge of disrupting the incumbent crystalline silicon technology. These efficiencies are a result of high optical absorption, long carrier diffusion lengths and excellent charge transport, and lead to the generation of exceptional open-circuit voltages (V_{oc}) as high as 1.2 V [2]. According to the Shockley-Queisser (SQ) limit, the photo-conversion efficiency (PCE) of PSCs with absorber band gaps of 1.6 eV can reach 30.14%, corresponding to a short circuit current density (J_{sc}) of 25.47 mA/cm^2, a V_{oc} of 1.309 V and a fill factor (FF) of 90.5% [3]. Fundamentally, a perovskite is any material which has a crystal structure that can be described by the general chemical formula ABX_3. Herein, A represents a cation, B represents a metal cation with two valence electrons, and X represents an anion [4]. Since a variety of elements can be chosen to fill the A, B or X locations in the crystal structure, perovskites can easily be tuned for their physical, optical, and electrical properties. The highest efficiencies reported to-date are for the organic–inorganic lead halide perovskite, $CH_3NH_3PbX_3$ (X = I, Br, Cl), for which the shorthand notation, $MAPbX_3$, is commonly used. Organic–inorganic lead halide perovskites will be the focus of this work, and, unless otherwise stated, the subject of any reference to the term perovskite.

Recent life cycle assessments and techno-economic analyses [5, 6], have indicated that delaying degradation and extending the lifetime of PSCs is essential for sustainability and commercial viability. Since competitive efficiencies have already been demonstrated, the success of PSCs relies now on the improvement of their stabilities. To ensure that this technology will be profitable, lifetimes of at least 15 years [6], but ideally 25 years should be realized [7]. The solution, however, is not so straightforward. Perovskites degrade readily upon exposure to oxygen and moisture, therefore necessitating strategies for degradation mitigation or prevention. Additionally, the perovskite crystals are thermally unstable and have low decomposition temperatures as a result of their ionic nature and the use of organic meythlammonium ($CH_3NH_3^+$, MA) cations. Photo-induced degradation of perovskites constitutes another major issue.

2. Perovskite degradation

Moisture is one of the most prevalent causes of degradation in PSCs. Water molecules that are able to permeate through the solar cell stack will react with the A-site organic cation in the perovskite and form hydrogen bonds. This weakens the bonds to the B- and X-site halogenated lead, rendering the perovskite more susceptible to thermal- and UV-induced degradation [8]. Additionally, water will react with X-site iodide ions to decompose the perovskite into hydroiodic acid (HI) and lead iodide (PbI_2) [8]. Therefore, to improve the intrinsic moisture stability of the perovskite, X-site and A-site substitutions have been suggested. For example, substituting X-site iodine with bromine increases the strength of cation-lead halide bonds, thereby reducing the susceptibility of the perovskite to moisture-induced degradation [9]. Further, since grain boundary defects act as a host for these detrimental reactions with water, passivating perovskite grain boundaries and increasing grain sizes has been found to extend perovskite lifetimes in humid environments [10, 11].

Oxygen is another significant contributor to degradation in PSCs. Oxidative degradation occurs significantly in both the charge transport and perovskite layers. Oxidation of organic charge transport layers results in compromised carrier mobility and solar conversion efficiencies [8]. Conversely, metal oxide charge transport layers (e.g., TiO_2, etc.) are not sensitive to oxidation, themselves. However, they can absorb oxygen, and when combined with UV light, photo-excitation yields reactive superoxide (O_2^-), which then catalyzes the rapid oxidative degradation of the adjacent perovskite layer [8]. Termed 'photo-oxidation,' this is an accelerated form of oxidative degradation, which occurs upon simultaneous exposure to UV light and oxygen. The perovskite crystal itself is also highly susceptible to photo-oxidation. Photo-excitation of the perovskite increases the density of halide vacancies, which serve as gateways for diffusion of oxygen into the perovskite lattice [12]. Again, superoxide species initiate the degradation, resulting in the formation of decomposition products such as yellow-colored PbI_2 [8]. Strategies such as doping the perovskite with cadmium (Cd) have been used to decrease the density of halide vacancies and increase intrinsic resistance to photo-oxidation [13]. Similarly, Gong et al. report a PSC with 10.4% efficiency that employs doping the perovskite with Se^{2-} to strengthen the interaction between the MA cation and its inorganic framework, thereby improving stability by 140 times compared to the undoped film, and achieving 70% PCE retention after 700 hours of exposure to air [14]. Further, since oxidative degradation is most harmful when catalyzed with UV light, filtering out energetic UV photons constitutes another promising strategy for extending device lifetimes.

While intrinsic stability improvements remain necessary to prevent water- and oxygen-induced degradation during manufacturing and assembly, encapsulation provides the most effective barrier against moisture and oxygen. Even so, since package leakage and small amounts of water and oxygen permeation are inevitable, enhancing intrinsic stability and encapsulating devices will likely need to be applied in synergy to provide sufficient protection from all catalysts of degradation. However, even when a hermetic encapsulation is achieved, (i.e., permeation of water and oxygen is considered negligible) PSCs still suffer from UV-induced degradation. For instance, illumination can result in the reversible segregation of halide and cation species, which can hinder the performance of devices [15].

Heat constitutes a final extrinsic stressor which can accelerate the reactions responsible for degradation in PSCs [8]. For example, the PbI_2 decomposition product has been observed from prolonged exposure of $MAPbI_3$ perovskites to temperatures as low as 85°C [16]. This can be detrimental since many manufacturing steps, including the annealing and encapsulating stages, occur at high temperature. Since organic materials are relatively volatile and are most sensitive to thermal degradation, the use of the common MA A-site organic cation can be problematic. Therefore, A-site cation substitution and mixing, with more thermally-stable materials such as formamidium (FA), cesium (Cs) and rubidium (Rb), is a popular strategy to improve thermal stability in perovskites [17]. All-inorganic PSCs represent another promising avenue towards stability, by eliminating issues associated with the thermal degradation of the organic cation. Liu et al. devised a $CsPbI_2Br$-based PSC with an efficiency of 13.3%, which exhibited 80% PCE retention after thermal treatment at 85°C for 360 hours [18]. Another stabilizing strategy was demonstrated by Yun et al., who to reduced photo- and thermal degradation by incorporating LiF passivators in organic–inorganic lead halide perovskites with efficiencies up to 20%. Remarkably, they observed 90% PCE retention after 1000 hours of exposure to 1 sun illumination or 85°C temperatures [19].

Significant progress has been on the intrinsic stabilization of PSCs. While the results are promising, no solution has been reported to-date that has demonstrated the long-term operation of PSCs in outdoor conditions. Therefore, it is clear that a *combination* of intrinsic stabilization and encapsulation strategies will be necessary to produce a PSC that can appropriately withstand the breadth of illumination, heat, moisture and oxygen conditions encountered during manufacturing and operation. Hereafter, this work will focus on reviewing the recent progress in PSC encapsulation and on introducing novel directions for further improvement.

3. Perovskite solar cell encapsulation

To promote commercial viability, a PSC encapsulation should: (1) be impermeable to water and oxygen; (2) prevent organic and halide materials volatized by illumination and/or heat as well as toxic lead-based degradation products from escaping into the environment; (3) have very high visible light transparency so as not to compromise device efficiencies, (4) be chemically inert, and; (4) have sufficient mechanical durability and abrasion-resistance to tolerate the stress, wear and weathering introduced in normal installation and operation. Important performance metrics for an encapsulant are the water vapor transmission rate (WVTR) and the oxygen transmission rate (OTR). These measurements quantify the amount of water vapor or oxygen that permeate through the encapsulation material per unit time. Since water and oxygen are two of the most pervasive sources of degradation in perovskites, these metrics give a good indication as to the overall quality of the encapsulation. An adequate seal is achieved when the WVTR and OTR are in the range of or less than 10^{-3}–10^{-6} $g{\bullet}m^{-2} \bullet day^{-1}$ and 10^{-4}–10^{-6} $cm^3{\bullet}m^{-2}{\bullet}day^{-1}{\bullet}atm^{-1}$, respectively [20, 21]. Additionally, though not typically a focus, the ideal encapsulation system should also provide protection against UV irradiation and act as a thermal barrier to prevent UV-induced and thermal degradation.

Generally, encapsulation strategies have involved either the deposition of a transparent thin film encapsulant or the use of an edge sealant material to encapsulate the device between sheets of glass or polymers [22]. This precedent provides a framework for diving the encapsulation techniques into the following categories: (1) glass-to-glass (Section 3.1); (2) polymer (Section 3.2), and; (3) inorganic thin film encapsulation (Section 3.3). In glass-to-glass encapsulation, a glass cover is used in conjunction with a sealant to form the protective packaging. Conversely, polymer encapsulation encompasses the strategies that employ polymeric barriers – either as cover sheets or thin films. Finally, in the third category, thin inorganic barrier films form the encapsulation. A fourth category – hybrid encapsulations – will also be introduced in Section 3.4, and consists of any combination of the aforementioned three encapsulation strategies.

Thin films (organic or inorganic) in particular are uniquely suitable for encapsulation because they can serve as dense, pin-hole-free barriers to oxygen and water, yet remain lightweight and thin enough to not adversely affect the mechanical flexibility of the solar stack and can thus be compatible with roll-to-roll processing. This work will briefly contextualize the progress made to-date in PSC encapsulation, with an emphasis on techniques that incorporate thin barrier films. The most noteworthy encapsulation examples from the literature are summarized in **Tables 1–4**. Therein, the PCE of the encapsulated PSC and a schematic of the encapsulation are provided. Additionally, the WVTR of the encapsulant and outcomes of stability testing (% PCE retained) are given to provide a framework for comparing encapsulation strategies.

3.1 Glass-to-glass encapsulation

Derived from the standard encapsulation technique of the silicon solar technology, glass-to-glass encapsulation sandwiches the PSC between two sheets of glass which are sealed together by means of a sealant. Since the WVTR and OTR of glass are near zero, glass-to-glass encapsulation provides excellent protection from water- and oxygen-induced degradation, while maintaining high light transparency. Furthermore, since glass is easy to clean, has very good mechanical durability and is currently more cost-effective than alternative encapsulating systems, it is considered a highly efficient and industrially attractive encapsulant material [8]. However, moisture and oxygen ingress through the sealant at the edges of glass-to-glass encapsulated devices is significant enough to cause degradation [35]. As a result, recent efforts have been placed on optimizing sealant materials such that the WVTR and OTR are minimized. For example, butyl rubber edge sealants, such as polyisobutylene (PIB), have attracted

Ref.	PCE	Encapsulation method	WVTR	Stability Test(s)	% PCE Retained
Glass-to-Glass Encapsulation					
Shi et al. [23]	9%	Glass with PIB seal 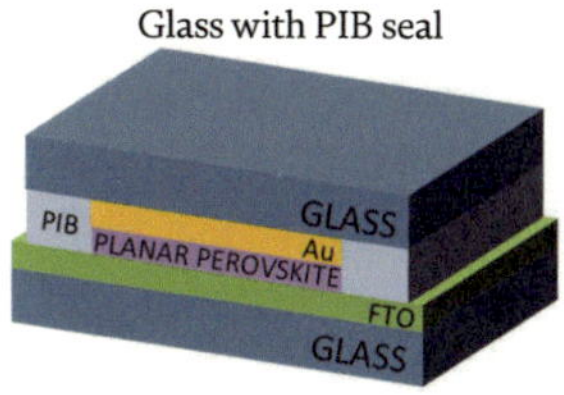	Glass: negligible PIB: 10^{-2}–10^{-3}	(1) 540 h at 85°C, 85% RH (2) 200 thermal cycles from −40–85°C	(1) ~ 100% (2) ~ 100%
Emami et al. [24]	8.2%	Glass frit 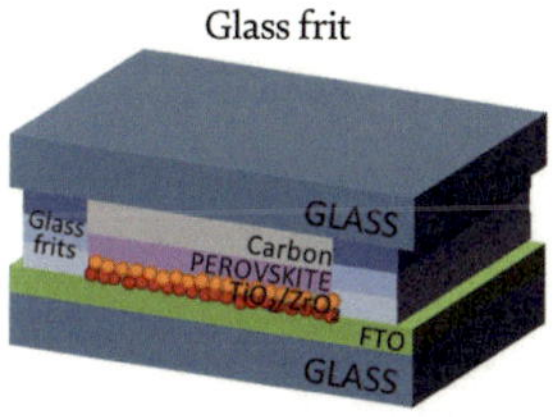	Hermetic	(1) 50 h at 85°C, 85% RH (2) 70 thermal cycles from −40–85°C	(1) ~ 100% (2) 98%

Table 1.
Notable glass-to-glass perovskite solar cell encapsulations from the literature. WVTR is reported in g·m^{-2}·day^{-1}.

considerable attention for their low WVTR (10^{-2}–10^{-3} g•m^{-2}•day^{-1}) [23]. Like PIB, many encapsulant adhesives and edge sealants are thermo-curable. However, curing at high temperature can degrade thermally unstable perovskites and reduce power conversion efficiencies even before aging tests begin [36]. While UV-curable epoxies are more costly, they are advantageous in that heat need not be applied to form the seal [35]. But, UV light, particularly in the presence of water and/or oxygen can also cause perovskite degradation. Nevertheless, Dong et al. observed a significant improvement in the PCE of devices encapsulated with a UV-curable epoxy (14.8%) compared to a thermally-curable one (8.9%) [32]. To eliminate the need for a sealant altogether, hermetic glass frit encapsulation has also been proposed [24]. **Table 1** compares PIB and glass frit sealed glass-to-glass encapsulations, demonstrating extremely low WVTR and high corresponding retained PCEs after aging. However, in both cases, the PSC has low initial PCE as a result of degradation caused by the encapsulation process and/or substitutions to internally stabilize the solar stack.

Many researchers believe that the competitiveness of PSCs lies almost exclusively in their efficiencies. Others are willing to incorporate more inexpensive materials and processes to reduce costs, even if it means sacrificing some efficiency. In order to keep the price per Watt ($/W) of a perovskite solar module low, these researchers are keen on retaining device flexibility, such that the solar cells can be made at the large-scale by low-cost roll-to-roll (R2R) processing. Recent work on ultra-thin glass encapsulation [37] has produced PSCs with retained flexibility, but further studies are required to properly assess their long-term stabilities.

3.2 Polymer encapsulation

Since glass-to-glass encapsulation is not inherently compatible with R2R processing, recent attention has been placed on polymer cover encapsulation. Herein, polymer sheets sealed with thermally-/UV-curable epoxies or pressure sensitive adhesives are used to

Ref.	PCE	Encapsulation method	WVTR	Stability test(s)	% PCE retained
Polymer Encapsulation					
Liu et al. [25]	8.6%	PDMS 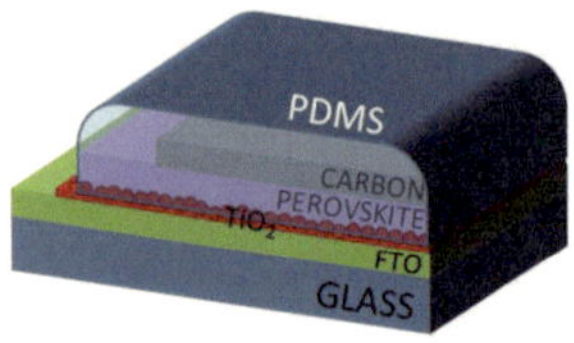	~10^2	3000 h at 25°C, 20% RH	~100%
Bella et al. [26]	19%	Downshifting fluoropolymer 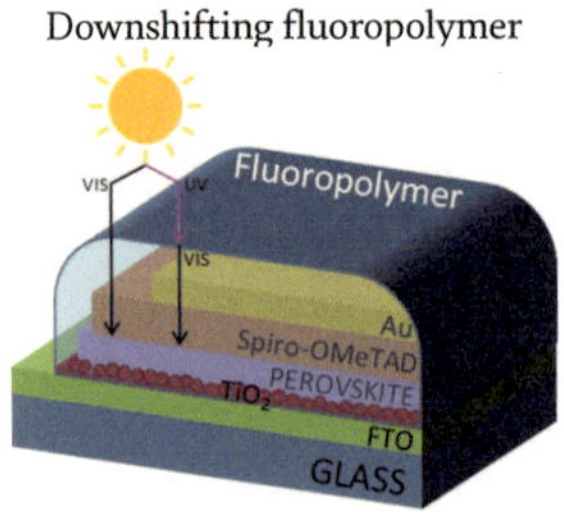	N.R.	(1) 3 months outdoors (rain, −3 to +27°C) (2) 1 month at 25°C, 95% RH	(1) 95% (2) 96%
Jang et al. [27]	17.2%	EVOH with SiO_2 and GO additives 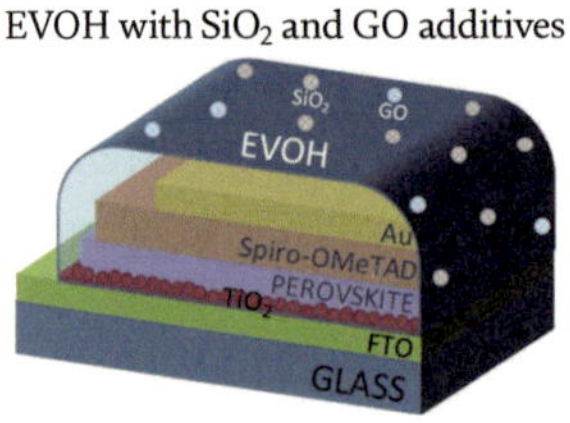	3.34×10^{-3}	5 h of direct contact with water at 25°C	86%

Table 2.
Notable polymer perovskite solar cell encapsulations from the literature. WVTR is reported at ambient conditions in $g{\cdot}m^{-2}{\cdot}day^{-1}$. 'N.R.' indicates that a value was 'not reported.'

Ref.	PCE	Encapsulation method	WVTR	Stability test(s)	% PCE retained
Inorganic Thin Film Encapsulation					
Choi et al. [28]	15%	Al_2O_3 by ALD at 95°C 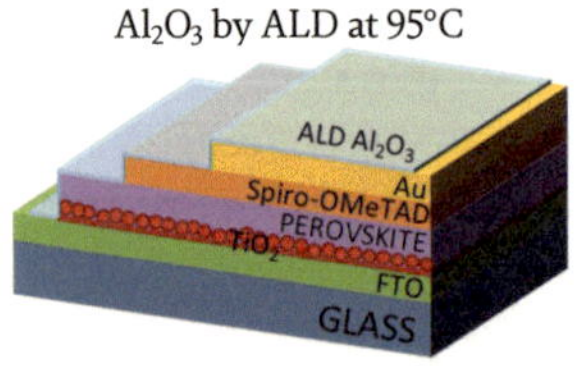	1.84×10^{-2} at 45°C, 85% RH	7500 h at 25°C, 50% RH	92%
Ramos et al. [29]	17.4%	Al_2O_3 by ALD at 60°C 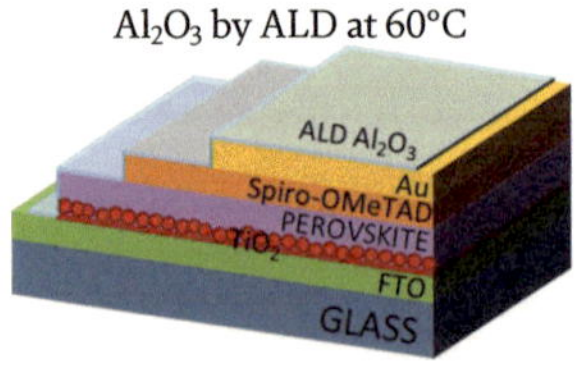	N.R.	2250 h at 25°C, 50% RH	75%

Table 3.
Notable thin film perovskite solar cell encapsulations from the literature. WVTR is reported in $g{\cdot}m^{-2}{\cdot}day^{-1}$. 'N.R.' indicates that a value was 'not reported.'

Ref.	PCE	Encapsulation method	WVTR	Stability test(s)	% PCE retained
Hybrid Encapsulation					
Chang et al. [30]	16.1%	Al_2O_3/PET 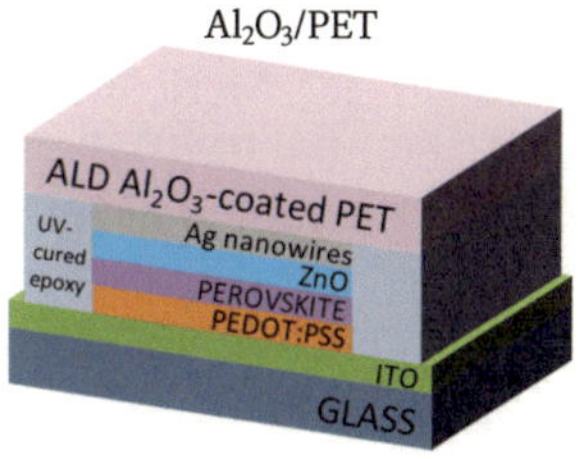	Al_2O_3: 9.0×10^{-4} at 50°C, 50% RH	42 days at 30°C, 65% RH	97%
Lee et al. [31]	18.5%	4-dyad Al_2O_3/pV3D3 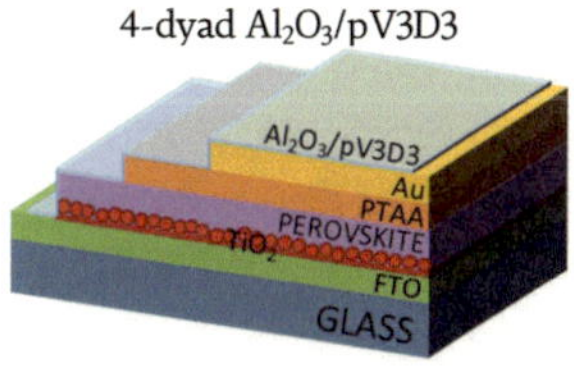	5.3×10^{-4} at 38°C, 90% RH	300 h at 50°C, 50% RH	97%
Dong et al. [32]	14.8%	SiO_2 with glass cover and desiccant sheet 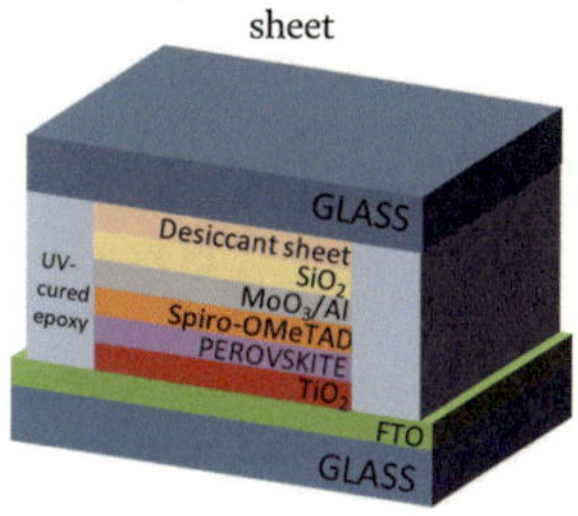	N.R.	(1) 48 h AM1.5 illumination at 85°C, 65% RH (2) 432 h outdoors (30–90% RH)	(1) 80% (2) ~ 100%
Liu et al. [33]	15.8%	Parylene with glass cover 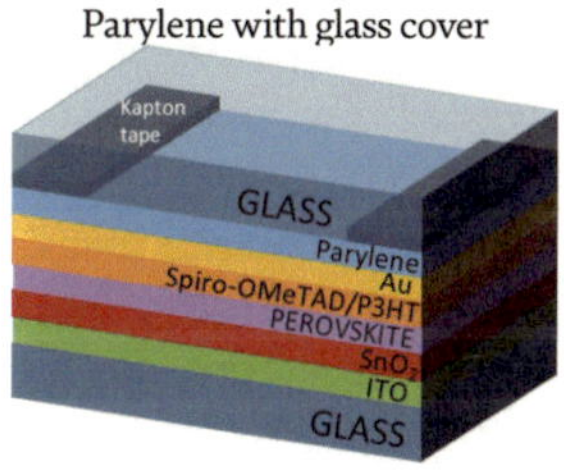	N.R.	2000 h continuous operation under AM1.5 illumination in N_2 environment with <5% RH	85%
Fumani et al. [34]	10%	Polymer resin (embedded with PEG) and glass substrate 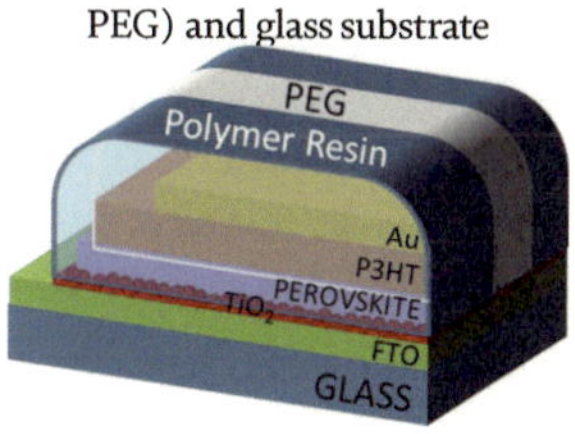	N.R.	830 days at 25°C, 28% RH	79%

Table 4.
Notable hybrid perovskite solar cell encapsulations from the literature. WVTR is reported in g·m^{-2}·day^{-1}. 'N.R.' indicates that a value was 'not reported.'

encapsulate a PSC, providing a barrier against extrinsic stressors such as moisture and oxygen. Many low-cost, flexible polymers, such as PET, PMMA and PC, have been used as encapsulating materials, however, much like with glass-to-glass encapsulation, moisture and oxygen ingress through sealants and degradation during curing persists [35]. To overcome this challenge, the deposition of solution-processed polymer layers directly on top of devices has been proposed [25, 38]. McKenna et al. [38] deposited 800 nm thin polymer films (i.e., PMMA, PC, EC, and PMP) directly on top of perovskite layers by spin coating and evaluated their ability to inhibit degradation. When exposed to 60°C heat at ambient conditions for 432 hours, the uncoated perovskite film was completely degraded, while the PMMA-encapsulated film remained in pristine condition (no evidence of PbI_2 formation). It is unsurprising that, of the polymers tested, PMMA provided the best device longevity because it has the lowest WVTR (55.2 $g{\bullet}m^{-2}{\bullet}day^{-1}$) and OTR (4.8 $cm^3{\bullet}m^{-2}{\bullet}day^{-1}{\bullet}atm^{-1}$) [38].

Bella et al. [26] also employed an innovative polymeric coating strategy to simultaneously slow water permeation and prevent UV-induced degradation of PSCs. By spin-coating devices with luminescent downshifting fluoropolymers that absorb incident UV light and re-emit it to the perovskite active layer as visible light, UV-induced degradation is effectively eliminated without sacrificing any photocurrent (PCE = 19%). Concurrently, the multifunctional polymeric coating is hydrophobic and provides a strong barrier to water-induced degradation. Notably, devices encapsulated and the top and bottom with ~5 μm films of this fluoropolymer demonstrated a 95% retention in PCE after 3 months of exposure to outdoor elements including heavy rain and temperatures ranging from −3 to +27°C.

Despite these promising findings, the relatively high WVTR (10^0–10^2 $g{\bullet}m^{-2}{\bullet}day^{-1}$) and OTR ($10^1$–$10^2$ $cm^3{\bullet}m^{-2}{\bullet}day^{-1}{\bullet}atm^{-1}$) of standard polymers limit long-term encapsulation effectiveness [35]. To this end, the incorporation of additives in polymer matrices to form polymer composite encapsulants with unique photo-, moisture- and/or oxygen-interactions has demonstrated potential for improved stability [27, 39, 40]. Jang et al. [27] fabricated a 100 μm-thick film of poly(vinyl alcohol-co-ethylene) (EVOH) copolymer with dispersions of SiO_2 and graphene oxide (GO) fillers. The SiO_2 inhibited water permeation by rendering the pathway for penetration through the polymer more tortuous, while the hydrophobicity of the GO repelled water molecules. By including these dispersions, the EVOH/SiO_2/GO composite polymer has a remarkable WVTR of 3.34×10^{-3} $g{\bullet}m^{-2}{\bullet}day^{-1}$, compared to 4.72×10^{-2} $g{\bullet}m^{-2}{\bullet}day^{-1}$ for EVOH only. PSCs encapsulated with the EVOH composite by means of a UV-curable adhesive retained 86% of their original PCE after 5 hours of direct exposure to water. **Table 2** summarizes the polymer encapsulation strategies discussed herein, demonstrating higher WVTR, on average, than with glass-to-glass encapsulations. While all stability tests yield high PCE retention, many of the test conditions were not as harsh as those described in **Table 1**, and constitute less accelerated forms of aging.

Inorganic materials such as metal oxides form denser films with substantially lower WVTR and OTR than their polymer counterparts. To take advantage of this, some researchers have combined transparent thin metal oxide films with polymer encapsulation to provide superior resistance to water- and oxygen-induced degradation. For example, Chang et al. [30] deposited 50 nm thin films of Al_2O_3 by ALD onto PET substrates that they then used to encapsulate PSC devices. The Al_2O_3 thin film served as an excellent barrier to moisture and oxygen, having WVTR and OTR of 9.0×10^{-4} $g{\bullet}m^{-2}{\bullet}day^{-1}$ and 1.9×10^{-3} $cm^3{\bullet}m^{-2}{\bullet}day^{-1}{\bullet}atm^{-1}$, respectively. Moderate increases in WVTR and OTR were observed for Al_2O_3-coated PET substrates subject to bend testing, indicating that while somewhat compatible with flexible devices, further effort may be required to increase the reliability and longevity of the encapsulation and to prevent partial delamination of rigid inorganic coatings from soft polymeric substrates. Nonetheless, encapsulated devices exposed to ambient conditions (30°C, 65% RH) for 42 days demonstrated negligible degradation in PCE.

The sequential combination of organic and inorganic layers to form organic–inorganic hybrid flexible multilayers has also been demonstrated to further reduce water and oxygen permeation through polymer-based encapsulants. The hybrid multilayers are deposited on polymer substrates, such as PET, that serve as the backbone for the encapsulation. Next, a series of organic polymer-based and metal-oxide inorganic thin films are deposited sequentially, wherein the organic layers help to retain flexibility and ductility and passivate interfacial defects, while the inorganic layers provide enhanced fortification against water and oxygen permeation [35]. WVTRs obtained at standard temperature and pressure (STP) for organic–inorganic hybrid multilayers are about three orders of magnitude less than that of uncoated PET; at elevated temperature and humidity (38°C, 90% RH), WVTR remains below 10^{-3} g•m^{-2}•day^{-1} [41]. Furthermore, deposition of these complex coating structures in roll-to-roll systems using vacuum-based techniques such as magnetron sputtering has established their compatibility with large-scale production. Kim et al. [42] combined the aforementioned benefits of organic–inorganic hybrid flexible multilayer coatings with the antireflective properties of Nb_2O_5/SiO_2/Nb_2O_5 thin films to create a protective barrier for PSCs that minimizes undesirable light reflection and enhances PCE (17%). Additional experimentation is required to assess the effect of these types of encapsulations on the long term stability of PSCs.

3.3 Inorganic thin film encapsulation

Thin film encapsulation (TFE), wherein a thin barrier film is deposited directly on top of the PSC, is considered a next-generation encapsulation strategy since it can overcome many of the issues associated with glass and/or polymer cover encapsulation (e.g., moisture ingress through edge seals). Importantly, TFE is simultaneously compatible with R2R processing and, depending on the deposition technique and material selected, ultra-low WVTR and OTR can be achieved. In fact, the large variety of materials (e.g., organics, inorganics, organic–inorganic composites, etc.) and deposition techniques (e.g., spin coating, CVD, PVD, etc.) available in TFE provides a unique opportunity to tailor the properties of the barrier coating to better suit the requirements of the application. Inorganic TFE is distinct from the polymeric thin film encapsulations discussed in Section 3.2 in that the thin barrier films are inorganic in nature. As previously detailed, these generally have the advantage of reduced WVTR and OTR compared to their polymeric thin film counterparts. However, a major concern of inorganic TFE is whether the deposition of the thin barrier film can be effectively and efficiently performed at a large-scale; many inorganic TFE strategies involve cost-prohibitive complex vacuum deposition systems with low deposition rates. Adhesion is another concern. Where the thermal expansion coefficient of the thin inorganic encapsulating film is substantially different than that of the solar stack, mechanical stress, stability testing and even normal operation may cause delamination.

Al_2O_3 has gained the most attention in inorganic TFE as a result of its high transparency, electrical insulation and extremely low WVTR (9.0×10^{-4} g•m^{-2}•day^{-1}) and OTR (1.9×10^{-3} cm^3•m^{-2}•day^{-1}•atm^{-1}) [28, 29, 43]. Atomic layer deposition (ALD) is often used to deposit the Al_2O_3 in TFE applications because of the high quality and uniformity of films produced [30]. However, as described in **Table 3**, a trade-off exists in selecting the ALD barrier-film deposition temperature. High temperature depositions yield pinhole/defect-free coatings, but can cause a significant decrease in PCE due to thermal degradation of organic materials during encapsulation. Conversely, low temperature depositions ensure that the thermally sensitive solar-stack retains high PCE after encapsulation, but yield Al_2O_3 films that are more prone to moisture and oxygen ingress. For example, while spiro-OMeTAD-based PSCs fabricated by Choi et al. [28] and encapsulated with 50 nm of ALD-deposited Al_2O_3 demonstrated excellent long-term stability in ambient environments (92% retention in PCE after 7500 hours at 25°C, 50% RH), the PCE of encapsulated devices was moderately compromised compared to that of un-encapsulated devices (16% drop in PCE after encapsulation), due to the elevated ALD deposition temperature of 95°C. Further increases in ALD deposition temperature lead to more severely compromised PCEs, particularly when organic hole transport materials (HTMs), such as spiro-OMeTAD, were used [28]. To prevent thermal degradation induced by the encapsulation process, deposition of Al_2O_3 by low-temperature ALD has been proposed. Ramos et al. [29] encapsulated spiro-OMETAD-based PSCs with 16 nm Al_2O_3 thin films deposited by ALD at 60°C. As a result of the reduced operating temperature, encapsulated PSCs had outstanding PCEs as high as 17.4%, representing a 93.6% retention of the original PCE, while the same cells encapsulated at 90°C exhibited a 54% loss in PCE. However, a higher defect density was observed in Al_2O_3 deposited at 60°C, leading to increased water permeation and worse long-term stability outcomes compared to high-temperature Al_2O_3 encapsulations. After 2250 hours of exposure to the same ambient conditions as in the previous study by Choi et al. (25°C, 50% RH), a more significant 25% drop in PCE was reported for PSCs encapsulated with the 60°C ALD-deposited Al_2O_3.

New and innovative strategies have been introduced to overcome the challenges associated with the deposition of a pinhole-free, low-temperature inorganic thin film for encapsulation. The inclusion of organic interlayers in the TFE constitutes one such proposition. The purpose of the organic barrier interlayers is to compensate for defects in the inorganic layers by elongating the pathway for water and oxygen permeation, effectively decreasing WVTR and OTR. Additionally, inorganic/organic encapsulations are less prone to delamination than their brittle, all-inorganic counterparts, due to a reduction in residual stresses and an improvement in flexibility [35]. Lee et al. [31] encapsulated PTAA-based PSCs with a 4-dyad multilayer stack of Al_2O_3 (21.5 nm)/pV3D3 (100 nm) deposited by ALD at 60°C and initiated chemical vapor deposition (iCVD) at 40°C, respectively. Low processing temperatures lead to negligible losses in PCE during encapsulation (<0.3%). Furthermore, the inclusion of pV3D3 organic interlayers produced PSCs with significantly improved stabilities in accelerated aging conditions: after storage for 300 hours at 50°C and 50% RH, PCEs retained 97% of their initial values. Importantly, this constitutes one of the best stabilities reported for a PSC with an encapsulated PCE higher than 18%.

While much has been done to advance TFE, lifetimes must be extended even further and harsher environmental testing is required to better assess their capacity for degradation prevention. Moreover, further optimization of the TFE deposition process is necessary to overcome limitations associated with the slow deposition rates, scalability and large operating costs of ALD.

4. Hybrid encapsulation

Glass-to-glass encapsulation remains one of most commercially-promising due to relatively low processing costs in conjunction with the effective seal produced.

However, moisture and oxygen ingress through epoxies and edge sealants remains a concern, preventing the realization of sufficiently long device lifetimes. Similarly, while polymer encapsulations are compatible with roll-to-roll processing, high WVTR and OTR limit effectiveness. Finally, while some early success has been demonstrated with TFE, much work remains to reduce operating costs and further improve lifetimes. It is therefore likely that a *hybrid* packaging which combines inorganic TFE with glass-to-glass or polymer encapsulation, thereby simultaneously taking advantage of the unique optical, mechanical and electronic properties of thin films materials and the strong barrier supplied by bulk polymers or glass, will provide the most effective and efficient means of preventing perovskite degradation.

Some of the most promising encapsulations that were previously described under other sub-sections actually employed combinations of glass-to-glass, polymer and inorganic thin film encapsulation and, as such, are more appropriately categorized as hybrid encapsulations. For example, the Al_2O_3-coated PET encapsulation reported by Chang et al. [30] that was first introduction in Section 3.2 on polymer encapsulation actually involves a combination of polymer and inorganic TFE strategies. It produced encapsulated devices that demonstrated negligible degradation in PCE after exposure to ambient conditions (30°C, 65% RH) for 42 days. Similarly, the PCE-enhancing organic–inorganic hybrid flexible multilayer coatings (PET/Nb_2O_5/SiO_2/Nb_2O_5/PPFC) by Kim et al. [42] were first introduced in Section 3.2 but more aptly constitute a hybrid encapsulation. Finally, Lee et al.'s 4-dyad multilayer stack of Al_2O_3/pV3D3 [31] described in the previous section combines polymer and thin film encapsulations to achieve remarkable PCE retention (97%) after storage for 300 hours at 50°C and 50% RH.

Also in pursuant with this hybrid strategy, Dong et al. [32] employed an encapsulation strategy wherein a 50 nm thin film of SiO_2 was deposited directly onto the device by electron beam deposition, followed by glass-to-glass encapsulation with UV-curable epoxy and a 180 μm piece of desiccant. Encapsulated devices were subject to accelerated aging tests and remarkably retained 80% of their original PCE after 48 hours under illumination at 85°C and 65% RH. Furthermore, almost full retention of PCE was reported after 432 hours of exposure to humid outdoor conditions where the relative humidity varied between 30 and 90%. Similarly, Liu et al. [33] encapsulated intrinsically stabilized PSCs with a 2 μm polymeric thin film of parylene by chemical vapor deposition and a cover glass. Impressively, by combining the principles of polymer, thin film and glass-to-glass encapsulation, the stability of the encapsulated devices under AM1.5 illumination was demonstrated for 2000 hours of continuous operation (PCE > 85% of initial value). Additionally, in one of the longest PSC stability tests published to-date, Fumani et al. [34] obtained 2-year stable PSCs by encapsulating the cathode and anode side of devices with 1.5 mm of polymer resin embedded with poly(ethylene glycol) (PEG) and glass, respectively. The resin provided a thick barrier against the diffusion of oxygen and water, and the PEG additive was used as a phase change material to limit device overheating cause by illumination. Freshly encapsulated devices had PCEs of 10%, which declined only slightly to 7.9% after 830 days (2.3 years) of storage in ambient conditions (25°C, 28% RH).

All of the hybrid encapsulation strategies described throughout this chapter are compared in **Table 4**. The diversity of all these hybrid encapsulations is reflected in the schematics and stability test results. Generally, hybrid encapsulations have allowed for high initial PCE and good PCE retention after aging. It should be noted that large differences in the severity of stability tests make direct comparison of different encapsulations difficult. A standardization of testing protocols would therefore provide a means for more efficient optimization of encapsulation techniques.

5. Conclusions & future research directions

Though stability of PSCs remains a concern, recent improvements to intrinsic stability and hybrid device encapsulation have produced PSCs with lifetimes up to 2 years [33, 34, 44]. Nonetheless, considerable progress remains to be made before the 15-year lifetimes required for economic feasibility are realized. Furthermore, many of the PSCs that have demonstrated long-term stability on the order of years are limited by relatively poor initial efficiencies due to material substitutions/eliminations for intrinsic stabilization.

As a result of the detrimental nature of water- and oxygen-induced degradation of PSCs, the majority of encapsulations focus on inhibiting moisture and oxygen ingress. However, limited work to-date has focused on using encapsulation strategies to target UV-induced degradation. In fact, UV light constitutes a major factor in perovskite degradation, not only because of illumination-induced reversible phase segregation, but because it catalyzes and accelerates moisture- and oxygen-induced degradation. Therefore, in the absence of UV light, it is conceivable to achieve sufficiently long PSC lifetimes, even for encapsulations with slightly higher-than-ideal WVTR and OTR. This opens the door to obtaining substantially better long-term stabilities with glass-to-glass encapsulation, where moisture and oxygen ingress through the edge sealant is somewhat inevitable, and even for polymer encapsulations, which are limited by inherently poor WVTR and OTR. Thus, the use of a thin film encapsulant material with optical properties tuned to screen or convert UV light into less energetic and harmful irradiation is very compelling. Future research should look to combine the benefits of glass-to-glass or polymer encapsulation with thin film UV-barriers to assess whether this constitutes a step towards PSC longevity. Nonetheless, one thing is for certain: the continued development and evolution of innovative encapsulation strategies such as this and all those presented in this work is certainly required to bridge the gap between lab-scale PSC success and large-scale commercialization.

Furthermore, this work has elucidated the difficulties associated with direct comparison of PSC encapsulations fabricated by different research groups due to the lack of consistency in aging and stability tests performed. In an effort towards test standardization, some researchers have performed PSC testing according to the International Electrotechnical Commission (IEC) standards (e.g., IEC61646), originally designed to assess the field performance of silicon photovoltaic modules. However, due to the differences in degradation pathways between silicon and perovskite photovoltaics, many researchers are critical that the IEC standards do not comprehensively appreciate or assess for all sources of degradation in PSCs. The testing standards proposed specifically for PSCs at the 2018 International Summit on Organic Photovoltaic Stability (ISOS) constitute a good starting point for discussions of PSC-specific stability tests [45]. However, future effort and consensus from the research community is still required to establish standardized testing protocols better suited to assess the long term stability of encapsulated PSCs.

Conflict of interest

Dongfang Yang is the Editor and Katherine Lochhead is Assistant to the Editor of this IntechOpen book: “Thin Film Deposition – Fundamentals, Processes and Applications.”

Author details

Katherine Lochhead[1*], Eric Johlin[2] and Dongfang Yang[1]

1 National Research Council Canada, London, ON, Canada

2 Western University, London, ON, Canada

*Address all correspondence to: katherine.lochhead@nrc-cnrc.gc.ca

References

[1] National Renewable Energy Laboratory (NREL). Best Research-Cell Efficiency Chart. 2022. Available from: https://www.nrel.gov/pv/cell-efficiency.html [Accessed: May 27, 2022]

[2] Liu Z, Kruckemeier L, Krogmeier B, Klingebiel B, Marquez JA, Levcenko S, et al. Open-circuit voltages exceeding 1.26 V in planar methylammonium lead iodide perovskite solar cells. ACS Energy Letters. 2019;**4**:110-117

[3] Park NG, Segawa H. Research direction toward theoretical efficiency in perovskite solar cells. ACS Photonics. 2018;**5**:2970-2977

[4] Park NG. Halide perovskite photovoltaics: History, progress, and perspectives. MRS Bulletin. 2018;**43**:527-533

[5] Gong J, Darling S, You F. Perovskite photovoltaics: Life-cycle assessment of energy and environmental impacts. Energy & Environmental Science. 2015;**8**:1953-1968

[6] Song Z, McElvany C, Phillips A, Celik I, Krantz P, Watthae S, et al. A technoeconomic analysis of perovskite solar module manufacturing with low-cost materials and techniques. Energy & Environmental Science. 2017;**10**:1297-1305

[7] Snaith HJ. Present status and future prospects of perovskite photovoltaics. Nature Materials. 2018;**10**:372-376

[8] Boyd C, Cheacharoen R, Leijtens T, McGehee M. Understanding degradation mechanisms and improving stability of perovskite photovoltaics. Chemical Reviews. 2019;**119**:3418-3451

[9] Noh JH, Im SH, Mandal TN, Seok SI. Chemical management of colourful, efficient, and stable inorganic-organic hybrid nanotructured solar cells. Nano Letters. 2013;**13**:1764-1769

[10] Yun JS, Kim J, Young T, Patterson RJ, Kim D, Seidel J, et al. Humidity-induced degradation via grain boundaries of $HC(NH_2)_2PbI_3$ planar perovskite solar cells. Advanced Functional Materials. 2018;**28**:1705363

[11] Wang Q, Chen B, Liu Y, Deng Y, Bai Y, Dong Q, et al. Scaling behaviour of moisture-induced grain degradation in polycrystalline hybrid perovskite thin films. Energy & Environmental Science. 2017;**10**:516-522

[12] Kim GY, Senocrate A, Yang TY, Gregori G, Gratzel M, Maier J. Large tunable photoeffect on ion conduction in halide perovskites and implications for photodecomposition. Nature Materials. 2018;**17**:445-449

[13] Saidaminov MI, Kim J, Jain A, Quintero-Bermudez R, Tan H, Long G, et al. Suppression of atomic vacancies via incorporation of isovalent small ions to increase the stability of halide perovskite solar cells in ambient air. Nature Energy. 2018;**3**:648-654

[14] Gong J, Yang M, Rebollar D, Rucinski J, Liveris Z, Zhu K, et al. Divalent anionic doping in perovskite solar cells for enhanced chemical stability. Advanced Materials. 2018;**30**:1800973

[15] Hoke E, Slotcavage D, Dohner E, Bowring A, Karunadasa HI, McGehee MD. Reversible photo-induced trap formation in mixed-halide hybrid perovskites for photovoltaics. Chemical Science. 2015;**6**:613-617

[16] Conings B, Drijkoningen J, Gauquelin N, Babayigit A, D'Haen J,

D'Olieslaeger L, et al. Intrinsic thermal instability of methylammonium lead trihalide perovskite. Advanced Energy Materials. 2015;**5**:1500477

[17] Turren-Cruz SH, Hagfeldt A, Saliba M. Methylammonium-free, high-performance, and stable perovskite solar cells on a planar architecture. Science. 2018;**362**:449-453

[18] Liu C, Li W, Zhang C, Ma Y, Fan J, Mai Y. All-inorganic $CsPbI_2Br$ perovskite solar cells with high efficiency exceeding 13%. Journal of the American Chemical Society. 2018;**140**:3825-3828

[19] Yun AJ, Kim J, Gil B, Woo H, Park K, Cho J, et al. Incorporation of lithium fluoride restraining thermal degradation and photodegradation of organometal halide perovskite solar cells. Energy & Environmental Science. 2020;**12**:50418-50425

[20] Griffini G, Turri S. Polymeric materials for long-term durability of photovoltaic systems. Journal of Applied Polymer Science. 2015;**133**:43080

[21] Cros S, de Bettignies R, Berson S, Bailly S, Maisse P, Lemaitre N, et al. Definition of encapsulation barrier requirements: A method applied to organic solar cells. Solar Energy Materials and Solar Cells. 2011;**95**:S65-S69

[22] Li J, Xia R, Qi W, Zhou X, Cheng J, Chen Y, et al. Encapsulation of perovskite solar cells for enhanced stability: Structures, materials and characterization. Journal of Power Sources. 2021;**48**:229313

[23] Shi L, Young TL, Kim J, Sheng Y, Wang L, Yifeng C, et al. Accelerated lifetime testing of organic-inorganic perovskite solar cells encapsulated by polysisobutylene. ACS Applied Material Interfaces. 2017;**9**:25073-25081

[24] Emami S, Martins J, Ivanou D, Mendes A. Advanced hermetic encapsulation of perovskite solar cells: The route to commercialization. Journal of Materials Chemistry A. 2020;**8**:2654-2662

[25] Liu Z, Sun B, Shi T, Tang Z, Liao G. Enhanced photovoltaic performance and stability of carbon counter electrode based perovskite solar cells encapsulated by PDMS. Journal of Materials Chemistry A. 2016;**4**:10700-10709

[26] Bella F, Griffini G, Correa-Baena JP, Saracco G, Gratzel M, Hagfeldt A, et al. Improving efficiency and stability of perovskite solar cells with photocurable fluropolymers. Science. 2016;**354**:203-206

[27] Jang JH, Kim BJ, Kim JH, Han E, Choi EY, Ji CH, et al. A novel approach for the development of moisture encapsulation poly(vinyl alcohol-co-ethylene) for perovskite solar cells. ACS Omega. 2019;**4**:9211-9218

[28] Choi EY, Kim J, Lim S, Han E, Ho-Baillie AWY, Park N. Enhancing stability for organic-inorganic perovskite solar cells by atomic layer deposited Al_2O_3 encapsulation. Solar Energy Materials and Solar Cells. 2018;**188**:37-45

[29] Ramos FJ, Maindron T, Bechu S, Rebai A, Fregnaux M, Bouttemy M, et al. Versatile perovskite solar cell encapsulation by low-temperature ALD-Al_2O_3 with long-term stability improvement. Sustainable Energy & Fuels. 2018;**2**:2468-2479

[30] Chang CY, Lee KT, Huang WK, Siao HY, Chang YC. High-performance, air-stable, low-temperature processed semitransparent perovskite solar cells enabled by atomic layer deposition. Chemistry of Materials. 2015;**27**:5122-5130

[31] Lee Y, Jeon NJ, Kim BJ, Shim H, Yang TY, Seok S, et al. A low-temperature

thin-film encapsulation for enhanced stability of a highly efficient perovskite solar cell. Advanced Energy Materials. 2018;**8**:1701928

[32] Dong Q, Liu F, Wong MK, Tam HW, Djurisic AB, Ng A, et al. Encapsulation of perovskite solar cells for high humidity conditions. ChemSusChem. 2016;**9**:2597-2603

[33] Liu Z, Qiu L, Ono LK, He S, Zhanhao H, Jiang M, et al. A holistic approach to interface stabilization for efficient perovskite solar modules with over 2,000-hour operational stability. Nature Energy. 2020;**5**:596-604

[34] Fumani NMR, Roghabadi FA, Maryam A, Sadrameli SM, Ahmadi V, Najafi F. Prolonged lifetime of perovskite solar cells using a moisture-blocked and temperature-controlled encapsulation system comprising a phase change material as a cooling agent. ACS Omega. 2020;**5**:7106-7114

[35] Corsini F, Griffini G. Recent progress in encapsulation strategies to enhance the stability of organometal halide perovskite solar cells. Journal of Physics: Energy. 2020;**2**:031002

[36] Aranda CA, Calio L, Salado M. Toward commercialization of stable devices: An overview on encapsulation of hybrid organic-inorganic perovskite solar cells. Crystals. 2021;**11**(519):2021

[37] Castro-Hermosa S, Lucarelli G, Top M, Hafland M, Fahlteich J, Brown TM. Perovskite photovoltaics on roll-to-roll coated ultra-thin glass as flexible high-efficiency indoor power generators. Cell Reports Physical Science. 2020;**1**:100045

[38] McKenna B, Troughton JR, Watson TM, Evans RC. Enhancing the stability of organolead halide perovskite films through polymer encapsulation. RSC Advances. 2017;**7**:32942-32951

[39] Fang HH, Yang J, Tao S, Adjokatse S, Kamminga ME, Ye J, et al. Unravelling light-induced degradation of layered perovskite crystals and design of efficient encapsulation for improved photostability. Advanced Functional Materials. 2018;**28**:1800305

[40] Han GS, Yoo JS, Yu F, Duff ML, Kang BK, Lee JK. Highly stable perovskite solar cells in humid and hot environments. Journal of Materials Chemistry A. 2017;**5**:14733-14740

[41] Nisato G, Klumbies H, Fahlteich J, Muller-Meskamp L, can de Weijer P, Bouten P, et al. Experimental comparison of high-performance water vapor permeation measurement methods. Organic Electronics. 2014;**15**:3746-3755

[42] Kim M, Kang TW, Kim SH, Jung EH, Park HH, Seo J, et al. Antireflective self-cleaning and protective film by continuous sputtering of a plasma polymer n inorganic multilayer for perovskite solar cells application. Solar Energy Materials and Solar Cells. 2019;**191**:55-61

[43] Singh R, Ghosh S, Subbiah AS, Mahuli N, Sarkar SK. ALD Al_2O_3 on hybrid perovskite solar cells: Unveiling the growth mechanism and long-term stability. Solar Energy Materials and Solar Cells. 2020;**205**:110289

[44] Grancini G, Roldan-Carmona C, Zimmermann I, Mosconi E, Lee X, Martineau D, et al. One-year stable perovskite solar cells by 2D/3D interface engineering. Nature. Communications. 2017;**815684**:1-8

[45] Khenkin MV, Katz EA, Abate A. Consensus statement for stability assessment and reporting for perovskite photovoltaics based on ISOS procedures. Nature Energy. 2020;**5**:35-49

Chapter 6

Spin Pumping in Magnetostrictive Ta/FeGaB/Ta Multilayer Thin Films

Karampuri Yadagiri and Tao Wu

Abstract

The magnetostrictive multilayer thin film stack (Ta/FeGaB(t)/Ta) deposited/ sputtered, studied the surface morphology, static and dynamic magnetic properties. The static magnetic properties multilayer studied; the coercive field and squareness increased for increasing thickness of FeGaB. The systematic study of damping in Ta/FeGaB/Ta multilayer performed by use of broadband ferromagnetic resonance (FMR) spectroscopy in-plan geometry in the range of temperature from 300 K to 100 K. The data were fitted to obtain the inhomogeneous line width (ΔH_0) and the damping factor (α). The damping factor is enhanced for the increased thickness of FeGaB. The enhancement of damping is due to spin pumping at the interface of Ta and FeGaB. The spin mixing conductance (g_{eff}) was calculated for magnetostriction thin films FeGaB; which had been increased for lowering the temperature. At 0 K, the g_{eff} of thin-film stack has 0.141×10^{18} m^{-2}. Therefore, the magnetostriction multilayer film stack can be used for magnonics, spin caloritronics, and spintronics applications.

Keywords: soft magnetic thin films, surface morphology of thin films, magnetization, ferromagnetic resonance, damping factor, spin pumping

1. Introduction

The recently, several phenomena/theories have been proposed across the interface of heavy metal/ferromagnetic (FM) such as the spin Hall effect (SHE) [1], spin-orbit torques [2], Dzyaloshinskii-Moriya interaction [3], and spin pumping [4]. In a while, the spin current mechanism has developed for non-magnetic materials (NM), which is one of the key points of modern spintronics. In the ferromagnet-nonmagnetic metal layer, a precessional magnetization in ferromagnetic layer generates as oscillating spin density, which can source a spin-polarized current to flow into the normal metal. This phenomenon is known as spin pumping [5–7], which offers the most interest of spin current. This spin current flows and dissipates into nonmagnetic metal, due to the influence of spin-orbit interaction. Subsequently, the damping factor enhances in NM/FM system [8–10].

Spin pumping theory [11] illustrates the relaxation of spin current in the NM layer, which denotes in way of spin-mixing conductance (g). The mixing conductance has been assumed as a properly of the NM. According to the theoretical model [12], spin pumping is a complex picture; however, it has been explained experimentally through the enhancement of damping factors for different materials. The effective mixing conductance (g_{eff}) consists of spin current across the interface of NM-FM, besides to the relaxation of spin current. The interface spin current characterizes as an effective specific interface spin resistance and relaxation associate with crossing the interface, called spin memory loss. The other model suggests the spin memory loss is due to interfacial spin-orbit interaction [10]. The interfacial spin resistance/spin memory loss, details of the FM-NM interface structure show an important role in determining the damping contribution due to spin pumping.

Magnetostrictive materials have been extensively utilized in vast applications such as sensors, actuators, micro-electrochemical-mechanical–systems (MEMS), and energy harvesters [13–16]. Among all the magnetostrictive materials, Terfenol-D has a large magnetostrictive constant (λ) 1600 ppm [17, 18], which is widely used in low-frequency devices, but the drawback of this material is hard to get saturation. The rare-earth free alloy, FeGa (Galfenol) shows great potential with high saturation magnetostriction of ~400 ppm for single crystal [19, 20] and ~280 ppm for directional solidified polycrystalline alloys [21]. It possesses a large saturation magnetization (~18 kG) at a low field (~100 Oe) [22]. However, the FeGa single-crystal films have been very lossy at microwave frequencies with a large line width of ferromagnetic resonance, which cannot be incorporated microwave magnetoelectric devices. The integration of metalloid element carbon into FeGa alloys are formed the $D0_3$ phase, which shows high saturation magnetostriction. This magnetostriction value is greater than that of FeGa binary alloys [22, 23]. Boron (B) is a well-known metalloid element, which is widely used in soft magnetic films for instance CoFeB thin films [24, 25]. Because of the B element inside CoFe thin film, the grain size of films refines and diminishes magneto crystalline anisotropy leading to excellent magnetic softness and microwave performance. In literature, the incorporation of B atoms in these FeGaB films can produce a nearly tripled saturation magnetostriction at a B content of 12 at.% [26]. The combination of soft magnetism, large magnetostriction constant, and excellent microwave magnetic properties make FeGaB film a potential candidate for magnetoelectric materials and other RF/microwave device applications. The magnetoelectric effects employ for creating electrostatically tunable [27] microwave resonators, phase shifters, and filters, which are important for applications in signal processing technologies [28, 29], and in schemes for performing logical processing operations using spin waves [30, 31].

The hybrid structure of ferromagnet (magnetostrictive ferromagnetic)/piezoelectric produce magnetoelectric effects [27]; which employs to construct efficient magnetic random access memory (MRAM) and spin-wave logical processing devices [32, 33]. The understanding of such nanoscale magnetoelectric devices requires the development of thin magnetic films with high magnetostriction constants. Therefore, the high value of magnetostriction of the film utilizes to reduce the line width and increase the magnitude of the magnetoelectric effect. The narrow resonant line widths and low damping are particularly important attributes of materials for microwave and spin-wave applications. So, we pick the FeGaB with 12 at.% of B; which has large magnetostriction constant and to investigate resonant linewidth and spin pumping across interface FeGaB and non-magnetic film.

In this work, we focus on the thickness dependence of magnetostrictive multilayer thin film stack (Ta/FeGaB(t)/Ta) deposition, surface morphology studies, static and dynamic magnetic properties. The thickness dependence of FMR shows the enhancement of the damping factor, which attributes the spin pumping across the interface of FeGaB and Ta. The spin-mixing conductance of the magnetostrictive multilayer thin film stack shows 0.081×10^{18} m^{-2} at 300 K, which is comparable with thin films of Si/SiO_2/Cu/Co(t)/Cu.

2. Experimental details

The tri-layer film stack (Ta/FeGaB/Ta) were deposited with DC & RF co-sputtering targets with base pressure < 1.0×10^{-7} Torr at room temperature onto fresh Si substrates. $Fe_{80}Ga_{20}$ and B targets are employed for FeGaB thin films. Here, the first Ta layer was employed as a buffer layer 25 nm, with different thickness (15, 25, 50, and 75 nm) of FeGaB film and 5 nm thick Ta film were deposited as the magnetic layer and the capping layer, respectively. The composition of the FeGaB film was characterized by using XPS and found as ~70, ~18, and ~12 at.% of Fe, Ga, and B, respectively. The surface morphologies images of samples were captured by Oxford Asylum MFP-3D AFM. The static magnetic properties of these samples were performed by employing Quantum Design® SQUID MPMS at room temperature. The dynamical magnetic properties studied using NanOsc Instruments Cryo-FMR in the VersaLab system with temperature variation 100–300 K and exciting frequency from 2 to 20 GHz.

3. Results and discussion

3.1 Surface morphology

Figure 1 shows the tri-layer film stack roughness is recorded by atomic force microscopy; all samples have shown roughness below 0.9 nm. This means that the high quality of films has been obtained.

3.2 Magnetic properties

Figure 2 shows the static magnetization of Ta (25)/FeGaB (t)/Ta (5). Magnetic moments have been increasing with increasing thickness of FeGaB. This means that magnetic spins of FeGaB are aligned along the easy axis. The magnetic coercive field (Hc) and squareness (Mr/Ms) have measured for the thickness of FeGaB film as shown in **Figure 2b**. As the thickness of film increases, Hc has initially decreased, in a while gradually increased. At film thickness, 25 nm has shown the low value of Hc

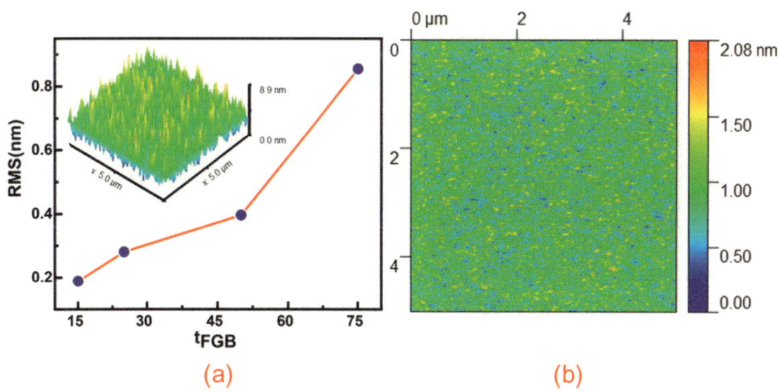

Figure 1
(a) The RMS values versus of thickness of FeGaB layer of thin film stack Ta (25)/FeGaB (15, 25, 50, and 75)/Ta (5); inset display the 3D image of thin films Ta (25)/FeGaB (75)/Ta (5). (b) AFM image of Ta (25)/FeGaB (15)/Ta (5).

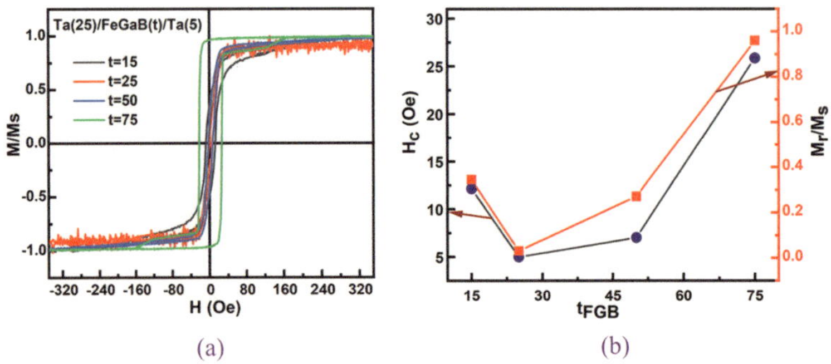

Figure 2.
(a) M-H loop of Ta (25)/FeGaB (t)/Ta (5) (b) coercive field (Hc) and squareness (Mr/Ms) of Ta (25)/FeGaB (t)/Ta versus thickness of FeGaB.

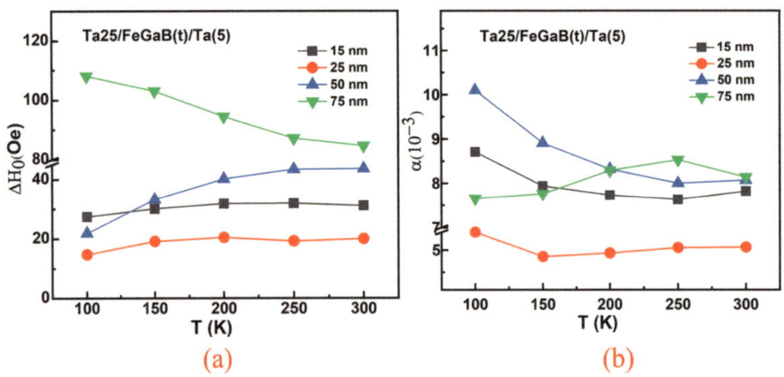

Figure 3.
(a) inhomogeneous of line width (ΔH_0) as a function of temperature (b) damping factor (α) versus temperature (T(K)) for Ta (25)/FeGaB (t)/Ta (5).

and squareness, due to the influence of the buffer layer on magnetic spins of FGB. The large value of Hc has been obtained for the 75 nm thickness of FGB.

3.3 Dynamic properties

Figure 3a and **b** shows the inhomogeneous of line width (ΔH_0) and damping factor (α) as a function of temperature. The ΔH_0 and α have been derived from the resonance line width (ΔH) – excitation frequency (GHz) by fitting $\Delta H = \Delta H_0 + \alpha f/\gamma$ [34, 35]. The thickness of FGB thin film increases, the (ΔH_0) is increasing at a specific temperature. The large thickness of FGB thin film (75 nm) has shown a large value of the ΔH_0 among all thin films. The temperature lowering to 100 K from 300 K, ΔH_0 is decreasing for FGB thin film thickness of 15–50 nm. Whereas, the 75 nm thickness of FGB thin film has shown an increasing trend, due to the creation of defects or structural imperfections for lowering the temperature.

Figure 3b shows the damping factor (α) as a function of temperature. The α has increased for lowering the temperature, except for the 75 nm thin film of FGB. The thickness of 50 nm FGB has shown a large damping value at 100 K. Because the 50 nm of FGB thin films shown a low value of ΔH_0 at 100 K. It means that the thin film has not produced any defects/structural imperfection, otherwise condensed the defects.

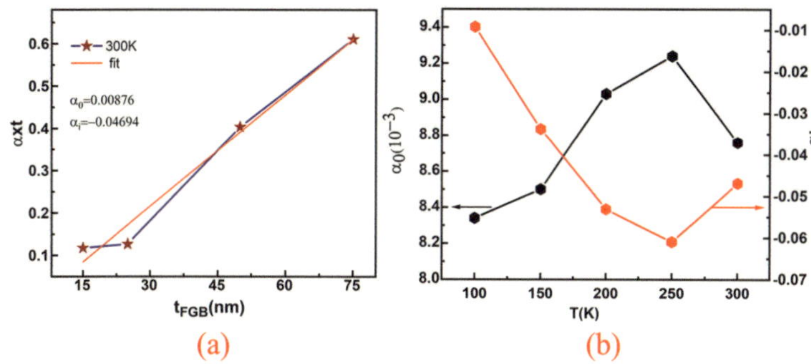

Figure 4.
(a) The thickness dependence of damping fitted with linearly to obtain surface/ inherent damping (α_o) and interface damping (α_i) (b) the plot of the obtained α_o and α_i temperature.

Therefore, the damping factor of the spin-wave is increasing for a 50 nm thin film of FGB. Whereas, 75 nm thickness FGB has exhibited a low value of damping due to multiple scattering of spin-wave from the defect and imperfections.

To find out interface and surface damping, damping factor (α) multiply with the thickness of FGB film and plotted with the thickness of FGB as shown in **Figure 4a**. The plot (αt vs. t) is linearly fitted an obtained the surface or inherent damping (α_0) and interface damping (α_i) parameters [35]. These parameters are varying for different temperatures. The surface damping is initially increasing for 300–250 K as shown in **Figure 4b**, later on, it is decreasing for lowering the temperature. Whereas the interface damping is exhibiting the opposite nature to surface damping. Finally, we can conclude that surface damping is mostly dominant at near to RT, and interface damping is dominant at a lower temperature in this film stack.

The effective magnetization has been found out from Kettle equation [35–37] fitted with frequency-resonance magnetic fields spectrum. Magnetic anisotropy (H_k) has been found along with effective magnetization, which shows a positive value and small value compare to effective magnetization. It means that no perpendicular anisotropy generates at the interface. Therefore, the effective magnetization is considering as surface magnetization (Ms). The parameter Ms. has been utilized for further calculation.

3.4 Spin pumping at the interface of Ta/ FGB

The spin pumping associate with the real part of the spin-mixing conductance (g_{eff}). The g parameter is proportional to the flux of angular momentum in the form of spin-polarized carriers the ferromagnet/nonmagnetic interface. This is seen by gyromagnetic precession in a ferromagnet. The enhanced damping factor has been found out by the subtraction of damping factor (α), inherent/surface damping (α_0). The inherent/surface, interface damping obtained from the plotting of damping-thickness of FGB as shown in **Figure 4b**. The enhanced damping is related to spin-mixing conductance and thickness of FGB [7, 11, 12, 38, 39].

$$\Delta\alpha = \alpha - \alpha_0 = g\mu_B \frac{g_{eff}}{4\pi M s t_{FGB}} \tag{1}$$

where g is Lande factor, μ_B is Bohr constant, Ms is saturation magnetization and t_{FGB} is the thickness of FeGaB magnetic film.

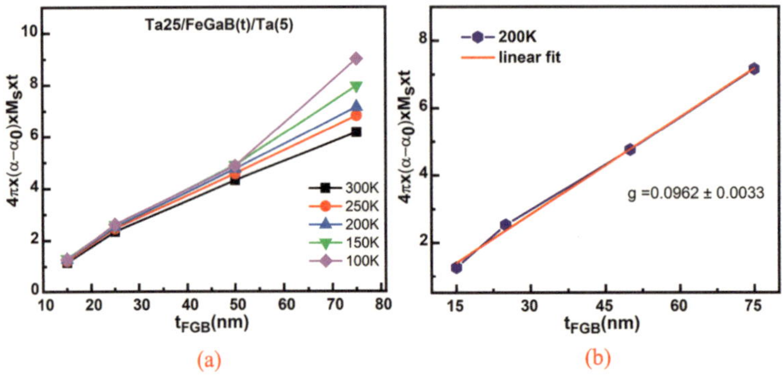

Figure 5.
(a) thickness depended on $4\pi \times (\alpha\_\alpha_o) \times Ms \times t$ *at different temperature 300–100 K for calculating the spin mixing conductance* (g_{eff}) *at the interface, (b) At 200 K, a linear fit of thickness depended on* $4\pi \times (\alpha\_\alpha_o) \times Ms \times t$ *provided the* g_{eff} *as* $0.0962 \times 10^{18}\ m^{-2}$.

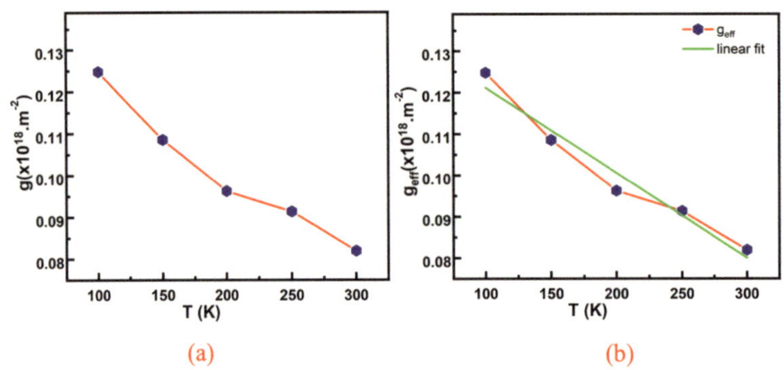

Figure 6.
(a) spin mixing conductance (g_{eff}) *as a function of temperature (T (K), (b) shows the linear fit of* g_{eff}*-T plot, the intercept of the plot provides spin mixing conductance at 0 K.*

To investigate the spin mixing conductance (g_{eff}), the enhanced damping factor has multiplied with saturation magnetization and thickness of magnetic films FeGaB. The product term has plotted with the thickness of FGB thin film, as shown in **Figure 5a**. The extraction of the spin mixing conductance (g_{eff}) has been obtained by linear fit as shown in **Figure 5b**. The mixing conductance value of the Ta/FeGaB/Ta thin film stack is $0.082 \times 10^{18}\ m^{-2}$ at 300 K, which has been increased gradually with lower temperature, as shown in **Figure 6**. This values are comparable with Si/SiO/Ta/Co(t)/Cu/Ta [40, 41], the Co/Cu films has mixing conductance $0.41 \times 10^{18}\ m^{-2}$. The conductance order is the same value; magnitude is different values. FeGaB thin film has a lower magnitude than that of the Co(t)/Cu. We can understand that spin pumping at the interface has reduced, due to the large thickness of FeGaB films and the large value of magnetostriction constant FeGaB.

Figure 6a shows the spin mixing conductance as a function of temperature. The mixing conductance has enhanced for lowering the temperature. At 100 K, the mixing conductance is a large value, $0.1245 \times 10^{18}\ m^{-2}$. It means that spin diffusion at the interface of Ta and FeGaB is increasing for low temperatures. We can conclude that spin mixing conductance depends on temperature. The linear fit of mixing conductance provides OK spin mixing conductance, as shown in **Figure 6b**. This value is $0.1417 \times 10^{18}\ m^{-2}$.

4. Conclusion

The multi-layer thin film stack Ta (25)/FeGaB (15,25,50 and 75)/Ta (5) deposited using the sputter. These film's surface morphology is found by AFM; the results suggested that good quality of thin films obtained. The static magnetization of the multi-layer thin film stack resulted as the magnetic moment increased for the increased thickness of FeGaB. The coercive field and squareness increased for a large thickness of FeGaB film. The dynamic magnetization of the multi-layer thin film stack informed that the inhomogeneous line width is increased for increasing the thickness of FeGaB; which is decreased for lowering the temperature. The damping factor decreased for lower temperatures. The thickness dependence of damping showed enhancement for increasing thickness of FeGaB and provided the inherent/surface and interface damping. The spin mixing conductance (g_{eff}) was calculated and increased for decreasing the temperature. The spin mixing conductance of Ta/FeGaB(t)/Ta is comparable with Co/Cu thin films. All results suggest the magnetostriction based thin film stack can be employed for magnonics, spin caloritronics, and spintronics applications.

Author details

Karampuri Yadagiri* and Tao Wu
School of Information Science and Technology, ShanghaiTech University, Shanghai, China

*Address all correspondence to: kistaiahyadagiri@gmail.com

References

[1] Wunderlich J, Kaestner B, Sinova J, Jungwirth T. Experimental observation of the spin-hall effect in a two-dimensional spin-orbit coupled semiconductor system. Physical Review Letters. 2005;**94**:047204

[2] Woo S, Mann M, Tan AJ, Caretta L, Beach GSD. Enhanced spin-orbit torques in Pt/Co/Ta heterostructures. Applied Physics Letters. 2014;**105**:212404

[3] Moriya T. Anisotropic superexchange interaction and weak ferromagnetism. Physical Review. 1960;**120**:91-98

[4] Brataas A, Tserkovnyak Y, Bauer GEW, Halperin BI. Spin battery operated by ferromagnetic resonance. Physical Review B. 2002;**66**:060404

[5] Tserkovnyak Y, Brataas A, Bauer GEW. Enhanced Gilbert damping in thin ferromagnetic films. Physical Review Letters. 2002;**88**:117601

[6] Mizukami S, Ando Y, Miyazaki T. Effect of spin diffusion on Gilbert damping for a very thin permalloy layer in Cu/permalloy/Cu/Pt films. Physical Review B. 2002;**66**:104413

[7] Mosendz O, Vlaminck V, Pearson JE, Fradin FY, Bauer GEW, Bader SD, et al. Detection and quantification of inverse spin Hall effect from spin pumping in permalloy/normal metal bilayers. Physical Review B. 2010;**82**:214403

[8] Qin H, Hämäläinen SJ, van Dijken S. Exchange-torque-induced excitation of perpendicular standing spin waves in nanometer-thick YIG films. Scientific Reports. 2018;**8**:5755

[9] Wang HL, Du CH, Pu Y, Adur R, Hammel PC, Yang FY. Scaling of spin Hall angle in 3d, 4d, and 5d metals from $Y_3Fe_5O_{12}$/metal spin pumping. Physical Review Letters. 2014;**112**:197201

[10] Chen K, Zhang S. Spin pumping in the presence of spin-orbit coupling. Physical Review Letters. 2015;**114**:126602

[11] Tserkovnyak Y, Brataas A, Bauer GEW, Halperin BI. Nonlocal magnetization dynamics in ferromagnetic heterostructures. Reviews of Modern Physics. 2005;**77**:1375-1421

[12] Liu Y, Yuan Z, Wesselink RJH, Starikov AA, Kelly PJ. Interface enhancement of Gilbert damping from first principles. Physical Review Letters. 2014;**113**:207202

[13] Hui Y, Nan TX, Sun NX, Rinaldi M. MEMS resonant magnetic field sensor based on an AlN/FeGaB bilayer nano-plate resonator. In: 2013 IEEE 26th International Conference on Micro Electro Mechanical Systems (MEMS). 2013. pp. 721-724

[14] Wu T, Chang C, Chung T, Carman G. Comparison of effective direct and converse magnetoelectric effects in laminate composites. IEEE Transactions on Magnetics. 2009;**45**:4333-4336

[15] Wu T, Bur A, Zhao P, Mohanchandra KP, Wong K, Wang KL, et al. Giant electric-field-induced reversible and permanent magnetization reorientation on magnetoelectric Ni/(011) [Pb(Mg1/3Nb2/3)O3](1−x)–[PbTiO3]x heterostructure. Applied Physics Letters. 2011;**98**:012504

[16] Wu T, Bur A, Wong K, Hockel JL, Hsu C-J, Kim HKD, et al. Electric-poling-induced magnetic anisotropy and electric-field-induced magnetization

reorientation in magnetoelectric Ni/(011) [Pb(Mg1/3Nb2/3)O3](1-x)-[PbTiO3]x heterostructure. Journal of Applied Physics. 2011;**109**:07D732

[17] Dong S, Li J-F, Viehland D. Vortex magnetic field sensor based on ring-type magnetoelectric laminate. Applied Physics Letters. 2004;**85**:2307-2309

[18] Dong S, Cheng J, Li JF, Viehland D. Enhanced magnetoelectric effects in laminate composites of Terfenol-D/Pb(Zr,Ti)O3 under resonant drive. Applied Physics Letters. 2003;**83**:4812-4814

[19] Clark A, Wun-Fogle M, Restorff JB, Lograsso TA. Magnetostrictive properties of Galfenol alloys under compressive stress. Materials Transactions. 2002;**43**:881-886

[20] Basantkumar RR, Stadler BJH, Robbins WP, Summers EM. Integration of thin-film Galfenol with MEMS cantilevers for magnetic actuation. IEEE Transactions on Magnetics. 2006;**42**:3102-3104

[21] Srisukhumbowornchai N, Guruswamy S. Large magnetostriction in directionally solidified FeGa and FeGaAl alloys. Journal of Applied Physics. 2001;**90**:5680-5688

[22] Clark AE, Restorff JB, Wun-Fogle M, Lograsso TA, Schlagel DL. Magnetostrictive properties of body-centered cubic Fe-Ga and Fe-Ga-Al alloys. IEEE Transactions on Magnetics. 2000;**36**:3238-3240

[23] Clark AE, Restorff JB, Wun-Fogle M, Hathaway KB, Lograsso TA, Huang M, et al. Magnetostriction of ternary Fe–Ga–X (X=C,V,Cr,Mn,Co,Rh) alloys. Journal of Applied Physics. 2007;**101**:09C507

[24] Platt CL, Minor NK, Klemmer TJ. Magnetic and structural properties of FeCoB thin films. IEEE Transactions on Magnetics. 2001;**37**:2302-2304

[25] Inyoung K, Jongryoul K, Ki Hyeon K, Yamaguchi M. Effects of boron contents on magnetic properties of Fe-Co-B thin films. IEEE Transactions on Magnetics. 2004;**40**:2706-2708

[26] Lou J, Insignares RE, Cai Z, Ziemer KS, Liu M, Sun NX. Soft magnetism, magnetostriction, and microwave properties of FeGaB thin films. Applied Physics Letters. 2007;**91**:182504

[27] Wu T, Bur A, Hockel JL, Wong K, Chung T, Carman GP. Electrical and mechanical manipulation of ferromagnetic properties in polycrystalline nickel thin film. IEEE Magnetics Letters. 2011;**2**:-6000104

[28] Fetisov YK, Srinivasan G. Electric field tuning characteristics of a ferrite-piezoelectric microwave resonator. Applied Physics Letters. 2006;**88**:143503

[29] Ustinov AB, Srinivasan G, Kalinikos BA. Ferrite-ferroelectric hybrid wave phase shifters. Applied Physics Letters. 2007;**90**:031913

[30] Schneider T, Serga AA, Leven B, Hillebrands B, Stamps RL, Kostylev MP. Realization of spin-wave logic gates. Applied Physics Letters. 2008;**92**:022505

[31] Ustinov AB, Lähderanta E, Inoue M, Kalinikos BA. Nonlinear spin-wave logic gates. IEEE Magnetics Letters. 2019;**10**:1-4

[32] Khitun A, Wang KL. Non-volatile magnonic logic circuits engineering. Journal of Applied Physics. 2011;**110**:034306

[33] Louis S, Lisenkov I, Nikitov S, Tyberkevych V, Slavin A. Bias-free spin-wave phase shifter for magnonic logic. AIP Advances. 2016;**6**:065103

[34] Kalarickal SS, Krivosik P, Wu M, Patton CE, Schneider ML, Kabos P, et al. Ferromagnetic resonance linewidth in metallic thin films: Comparison of measurement methods. Journal of Applied Physics. 2006;**99**:093909

[35] Yadagiri K, Wu T. The thickness of buffer layer and temperature dependent magneto dynamic properties of Ta/FeGaB/Ta tri-layer. Journal of Magnetism and Magnetic Materials. 2020;**515**:167277

[36] Herring C, Kittel C. On the theory of spin waves in ferromagnetic media. Physical Review. 1951;**81**:869-880

[37] Kittel C. Interpretation of anomalous Larmor frequencies in ferromagnetic resonance experiment. Physical Review. 1947;**71**:270-271

[38] Barati E, Cinal M, Edwards DM, Umerski A. Gilbert damping in magnetic layered systems. Physical Review B. 2014;**90**:014420

[39] Shaw JM, Nembach HT, Silva TJ. Determination of spin pumping as a source of linewidth in sputtered $Co_{90}Fe_{10}$/Pd multilayers by use of broadband ferromagnetic resonance spectroscopy. Physical Review B. 2012;**85**:054412

[40] Beaujour JML, Lee JH, Kent AD, Krycka K, Kao CC. Magnetization damping in ultrathin polycrystalline Co films: Evidence for nonlocal effects. Physical Review B. 2006;**74**:214405

[41] Tokaç M, Bunyaev SA, Kakazei GN, Schmool DS, Atkinson D, Hindmarch AT. Interfacial structure dependent spin mixing conductance in cobalt thin films. Physical Review Letters. 2015;**115**:056601

Chapter 7

Photoactive Heterostructures Based on α-Fe_2O_3 and CuO Thin Films for the Removal of Pollutants from Aqueous Solutions

Elizabeth C. Pastrana, Pierre G. Ramos, Luis A. Sánchez and Juan M. Rodriguez

Abstract

Heterostructured photoactive nanomaterials represent innovative construction to absorb UV and UV-vis light. This feature makes heterostructures exciting candidates for environmental photocatalytic applications such as organic pollutants degradation and removal of heavy metals, among others. Therefore, the efficient design of heterostructures based on thin films of oxide semiconductors will allow obtaining a novel material with outstanding properties. This work presents a review of the current heterostructures based on α-Fe_2O_3 and CuO thin films, which were deposited onto different substrates using physics and chemistry routes. Moreover, we will discuss the key factors to promote structural and morphology control and the drawbacks such as low absorption of the solar spectra, low active surface area, and charge carrier recombination. Finally, the relevance of the results and future directions of the heterostructures as materials for the purification of aqueous systems were discussed.

Keywords: heterostructure, semiconductor, α-Fe_2O_3, CuO, photoelectrochemical, photocatalysis, bandgap

1. Introduction

The heterostructures represent innovative constructions that use a broad s pectrum of sunlight, capable of efficiently absorbing UV and visible light. Besides, due to their response to the absorption of different light sources, heterostructures can be used for varied photocatalytic applications for environmental remediation such as water splitting, CO_2 conversion, photocatalytic degradation, and oxidation [1]. Heterostructures are composed of two or more semiconductor material structures with specific chemical compositions, which can be formed by an interface between two different materials with unequal bandgaps. Remarkably, the idea of combining various metal oxides to form heterostructures is relatively recent and was born as a response to the need to improve its morphological, structural, and functional properties [2, 3]. The development of semiconductor heterostructures has brought about a tremendous impact on our lives. The utilization of these devices has improved our quality of life due to their role in electronics, memory devices, photodetectors, and optoelectronic devices [4]. Transistors, photovoltaic cells, diodes, and sensors are some cases where heterostructures are present.

A heterostructure is essentially a physical, and therefore electronic, the bond between two different materials in the solid state [4, 5]. Whenever materials connect as a heterostructure, the Fermi levels (E_F) align: higher-energy electrons flow across the interface to lower-energy unoccupied states until the Fermi levels have equilibrated. This leads to creating a charge carrier depletion zone at the interface (depletion region). After that, an energy barrier potential is created at the interface due to band bending, caused by the difference in the initial Fermi levels of the materials that make up the heterostructure. Thus, charge carriers must overcome this potential energy barrier to cross the interface [6, 7]. The junction between two different materials is one of the most important aspects to consider in the behavior of the new heterostructured material [8].

Heterostructures can be classified according to the configuration and dimensions of the interface between the two components that conform: one-dimensional (1D), two-dimensional (2D), and three-dimensional (3D) heterostructures [9–11]. In a 1D heterostructure, the interface is similar to a line shape, being the contact area is only in one direction; usually, some 1D heterostructures are nanorods, nanotubes, nanofibers, and others [12]. 2D heterostructures have an interface plane-like shape and conform to layers for both components ("layer-by-layer" systems). These heterostructures (nanofilms) usually are formed on a substrate that can be conductive or not [13]. Finally, 3D heterostructures expand in all three directions and involve multiple and rare shapes, the most common considering nanoparticle components. Frequently, these heterostructures are conformed by agglomerates of crystallites of different components, the more popular is the core-shell structure [14]. There is a growing interest in the research for nanofilms (2D) materials with novel properties that permit stacking and combination of thin layer-layer reaching unexpected features such as technological goals [15]. These materials have an impressive performance concerning flexibility in their electronic role and convenient design. The thin film nanoheterostructures have a principal advantage: an efficient charge separation, which restricts the recombination of charge carriers and consequently increments their photoactivity performance [16]. However, they have a disadvantage: the loss of energy by the charge carriers, inhibiting the evolution of chemical reactions.

The nanosized thin films combined into heterostructures depend on purposes and the demanded functions. Nevertheless, the heterostructured photoactive materials are extensive and diversified; most attention is given to metal-semiconductor and semiconductor-semiconductor heterostructures [17]. In this line of thought, we focus on thin film semiconductor-semiconductor systems with interesting optical properties for various applications in advanced catalytic and healthcare fields.

A group of those nanoheterostructured photoactive materials can be presented in three different assemblies: hosted nanophase, when one phase grows onto another in many positions; segmented nanophase, which includes two materials within each particle; and a mixture of two nanophase materials [11]. Up to now, there have been many original articles with methodologies appropriate for heterostructures thin film fabrication. For example, heterostructure thin films could be assembled using electrodes and are easily routed for deposited films. On the other hand, inorganic thin films commonly use a wet-chemical bottom-up became to achieve worldwide attention, leading to a remarkable increase in the number of research papers and patents.

Notwithstanding, there have been very few researches focusing on various solution-processes techniques used for optimum inorganic nanofilm fabrication [18].

Metal oxides have a wide range of crystalline structures and a variety of functional properties that convert into unattainable conventional semiconductors. Iron oxide is one of the most abundant compounds on the planet, and it is a low-cost and environmentally friendly resource [19]. As a nanomaterial has outstanding characteristics that include; (i) photocatalytic properties for conversion reactions; (ii) large energy storage capacity; (iii) reduced industrial process cost; and iv) long-term sustainability due to its availability and low toxicity, along with others [20]. In particular, hematite (α-Fe_2O_3) is an n-type semiconductor with a bandgap of 1.9–2.2 eV, which ensures the absorption of more than half of visible light (>600 nm), obtaining 40% of the incident energy of the solar spectrum [21, 22]. On the other hand, copper oxide (CuO) is a p-type semiconductor with a bandgap from 1.3 to 2.2 eV. It has a monoclinic structure with fascinating characteristics: super thermal conductivity, high solar absorbance, low thermal emittance, relatively good electrical properties, photovoltaic properties, high stability, and antimicrobial activity [23, 24]. This semiconductor is involved in many technological fields, for instance, catalysis, sensors, high-efficiency thermal conductivity material, magnetic recording media, selectivity, and solar cell applications [25].

Based on the abovementioned, in this chapter, we attempt to provide the readers with an essential introduction to the innovative construction of heterostructured photoactive nanomaterials. We focus on the designs of heterostructures based on α-Fe_2O_3 and CuO thin films, highlighting the recent advances in heterostructures fabricated using different physics and chemistry routes. Finally, the progress in constructing thin film heterostructures as environmental technologies for the remediation of harmful elements in aqueous systems was presented.

2. Heterostructures based on α-Fe_2O_3

Hematite (α-Fe_2O_3) is a typical n-type semiconductor that has been extensively researched in photoelectrochemical and photocatalytic fields due to its being earth-abundant, cheap, and high chemical stability [26–28]. In addition, it has a bandgap energy of 1.9–2.2 eV that allows increased absorption of visible light [21]. Nevertheless, the photocatalytic performance of α-Fe_2O_3 is adversely affected by certain intrinsic factors such as the high recombination rate of charge carriers, poor conductivity of majority carriers, the short diffusion length of minority carriers, and a short lifetime of the electron-hole pairs [29]. Therefore, to overcome these drawbacks, many efforts have been developed, especially the formation of α-Fe_2O_3 heterostructures with other semiconducting materials [30–32]. The results revealed the successful construction of new materials with high photocatalytic efficiency in the degradation of various pollutants in wastewater [31–34].

For instance, Fe_2O_3/ZnO heterostructures thin films are one of the best materials used as photocatalysts because of their band position and bandgap energy of Fe_2O_3 and ZnO, which one is low bandgap energy lying in the visible region (Fe_2O_3), while the other in the UV region due to large bandgap (ZnO). Yu et al. [35] synthesized heterostructured Fe_2O_3-ZnO films, onto alumina plates, by a deposition method named "Solution Precursor Plasma Spray." The heterostructures films with different architectures were prepared by different injection modes. The first architecture, Fe_2O_3-ZnO-M, was obtained from the deposit of 12 layers of a mixed solution of the precursors of iron and zinc. Meanwhile, the samples labeled Fe_2O_3-ZnO-S3, Fe_2O_3-ZnO-S6, and Fe_2O_3-ZnO-S12 were prepared alternately, injecting zinc and iron precursor solutions with stipulated cycles. The layers deposited per cycle for the Fe_2O_3-ZnO-S3, Fe_2O_3-ZnO-S6, and Fe_2O_3-ZnO-S12 thin films were 3, 6, and 12, respectively.

The optical bandgap for the different Fe_2O_3-ZnO film architectures was obtained from the Kubelka-Munk function and ranged from 2.65 eV to 2.93 eV (**Figure 1a**). Although it is true that from the point of view of photon energy absorption, a higher photocatalytic activity under visible light irradiation will be obtained for the samples fabricated from the mixture of the precursor solutions, there are other parameters such as the morphology of the surface, which indicates that the samples prepared via the separated-injection mode can also enhance the photoactivity. Thus, the influence of these parameters on photocatalytic activity will be compared and discussed later.

The photocatalytic activities of Fe_2O_3-ZnO thin films were evaluated to degrade the Acid Orange 7 (Orange II) dye under UV and visible light irradiation. **Figure 1b** shows the photocatalytic degradation curves of the dye under UV light irradiation for Fe_2O_3-ZnO samples. The photodegradation efficiencies obtained follow the order: Fe_2O_3-ZnO-M (100%) > Fe_2O_3-ZnO-S12 (83%) > Fe_2O_3-ZnO-S3 (65%) > Fe_2O_3-ZnO-S6 (21%). Meanwhile, the corresponding photodegradation curves for the Fe_2O_3-ZnO samples under visible light irradiation are shown in **Figure 1c** and confirm that up to 95% degradation was achieved using the Fe_2O_3-ZnO-M film, followed by Fe_2O_3-ZnO-S12 (28.4%), Fe_2O_3-ZnO-S3 (22%) and Fe_2O_3-ZnO-S6 (15%). This order is the same as that obtained under UV light irradiation (**Figure 1b**).

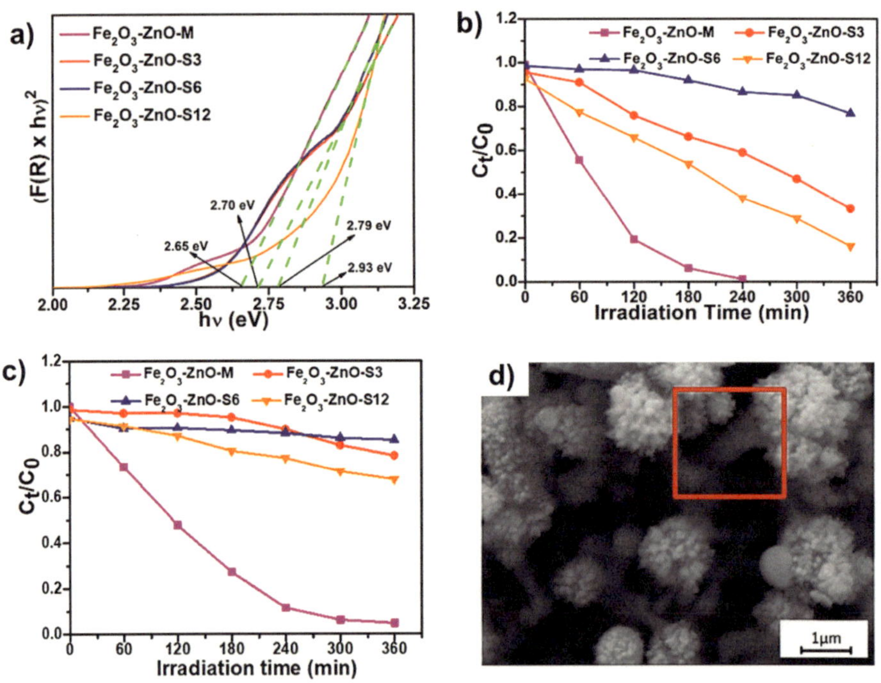

Figure 1.
(a) Kubelka-Munk plots for Fe_2O_3-ZnO films; (b) Photodegradation performances of the Orange II degradation under UV light irradiation for Fe_2O_3-ZnO films; (c) Photocatalytic test of Fe_2O_3-ZnO films photocatalysts for Orange II dye degradation under visible-light illumination; (d) FESEM images of the top view of Fe_2O_3-ZnO-S12 samples. Reproduced with permission from [35].

The surface morphologies influence the different photodegradation performances for the obtained samples under UV and visible light irradiation [35]. For example, the good photocatalytic performance obtained for the Fe_2O_3-ZnO-S12 sample is related to its well-shaped rod-like structure and hierarchical microstructure, as shown in the marked red rectangle in **Figure 1d**. Indeed, it is well known that the hierarchical structures present a highly organized structure, which allows an increase in the rate of mass transfer for reactant adsorption, increases the specific surface, and strongly favors efficient harvesting of light [36]. Thus, enhancing the photocatalytic properties. Furthermore, the high photodegradation efficiency of the Fe_2O_3-ZnO-M sample is attributed to the synergistic effect of ZnO and Fe_2O_3 phases to form a heterostructured catalyst, which allows a better separation of photogenerated electron/hole pairs, and thus a better photodegradation performance.

In addition, Suryavanshi et al. [37, 38] successfully synthesized stratified Fe_2O_3/ZnO thin films onto FTO coated glass substrate by chemical spray pyrolysis technique for subsequent study as photoelectrode in photoelectrocatalytic degradation of benzoic acid (BA), salicylic acid (SA) and methyl orange (MO) under solar illumination. The FE-SEM image of Fe_2O_3/ZnO heterostructure thin film is shown in **Figure 2a**. A morphology composed of grains randomly distributed on the film's surface was observed, which was helpful in the case of photocatalysis for the degradation of organic pollutants. Besides, the heterostructure morphology was slightly different from bare Fe_2O_3 and ZnO thin films. The variation of the morphologies for the thin films could be related to the lattice structure and defects produced during the deposition of films and the subsequent nucleation and growth [39]. Besides, the cross-section image of Fe_2O_3/ZnO film shown in **Figure 2b** reveals that the thickness obtained was about 2.04 μm.

The photoelectrocatalytic degradation experiment was carried out using a photoelectrochemical reactor model under the solar light illumination to investigate the photocatalytic activity of a large area deposited Fe_2O_3/ZnO photoelectrode. A constant bias (1.6 eV) was applied during the experiment to decrease the electron-hole recombination and increase the electrochemical reaction rate. During the experiments, it is observed that the concentration of pollutants decreases due to their photoelectrochemical oxidation. **Figure 3a-c** show the extinction spectra of benzoic acid, salicylic acid, and methyl orange for heterostructured Fe_2O_3/ZnO photoelectrode, respectively. It is observed that extinction peak intensity decreases over time due to the decomposition of pollutants. The inset plot in **Figure 3a-c** show the degradation efficiency versus reaction time of BA, SA, and MO dye for Fe_2O_3/

Figure 2.
FESEM image of (a) top and (b) cross-section view of stratified Fe_2O_3/ZnO thin films. Reproduced with permission from [38].

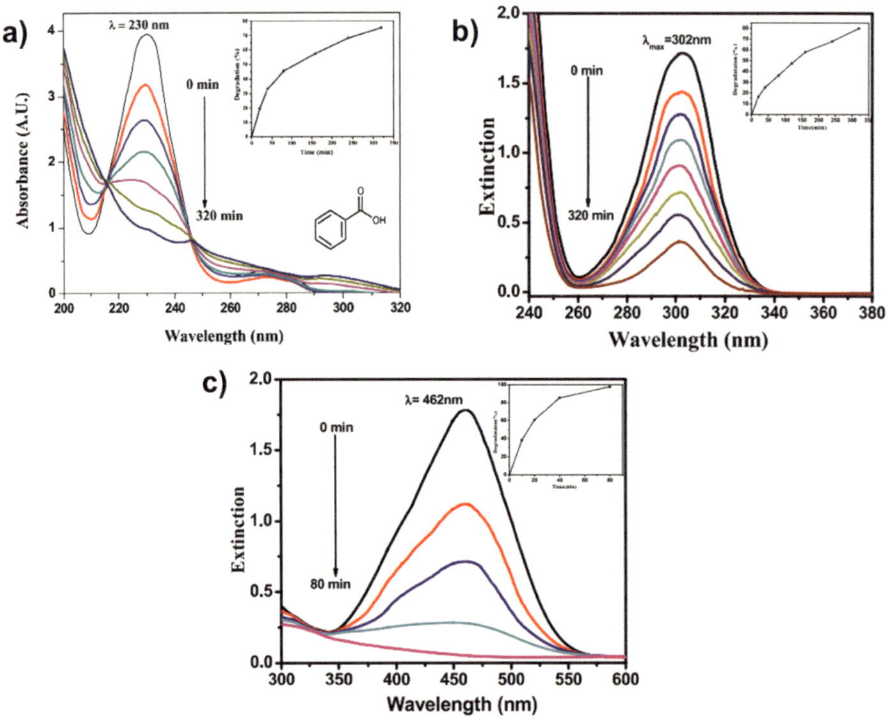

Figure 3.
Extinction spectra of (a) BA; (b) SA, and (c) MO as a function of wavelength at various time intervals. The percentage degradation using the photoelectrocatalytic degradation process is shown as insets. Reproduced with permission from [37, 38].

ZnO photoelectrode. Degradation efficiency (%) of pollutants was calculated using the following Eq. (1) [40]:

$$\text{Degradation efficiency}\,(\%) = \left(\frac{C_0 - C_t}{C_0}\right) \times 100 \tag{1}$$

Where, C_0 represents absorbance at t = 0 min, C_t is the absorbance at time t (reaction time). As shown in **Figure 3**, the prominent absorption peaks of BA, SA, and MO dye observed at 230 nm, 302 nm, and 462 nm, respectively, decrease continuously concerning the reaction time using stratified Fe_2O_3/ZnO photoelectrode. The degradation graph shows that there is 74% degradation of BA, 80% degradation of SA, and 98% of MO dye in 320 min for the first two and in 80 min for the dye, respectively.

The higher degradation efficiency for the heterostructure Fe_2O_3/ZnO occurs due to the photogenerated electron and holes can easily transfer from the conduction band (CB) of Fe_2O_3 to the CB of ZnO and from the valence band (VB) of ZnO to the VB of Fe_2O_3, respectively. This charge transfer is possible due to the valence and conduction band of Fe_2O_3 being located at higher positive potentials than that of ZnO. As a result, more separation of photogenerated electron-hole pairs takes place improving the photocatalytic performance. Then, on the surface of the photocatalyst, photogenerated holes directly interact with adsorbed H_2O onto the surface of the photoelectrode to produce a large amount of hydroxyl ($\bullet OH$) radicals; meanwhile, electrons transfer towards the counter electrode where they can interact with the oxygen present in the atmosphere to produce oxygen radical anions (O_2^-). Finally, the resulting ($\bullet OH$) radicals can react with organic pollutants present in the wastewater to degrade them into less harmful products such as H_2O and CO_2 [41].

TiO_2 is another semiconductor used as a coupling for Fe_2O_3-based photocatalysts to improve their photocatalytic activity. Wannapop et al. [42] fabricated Fe_2O_3/TiO_2 thin films for degradation of Rhodamine B (RhB) dissolved in water under UV light irradiation. In this research, photoluminescence (PL) measurements were carried out to evaluate the separation and recombination of photogenerated carriers, wherein a stronger emitted signal is related to a faster electron-hole recombination rate [42, 43]. The PL spectra obtained from the samples are shown in **Figure 4a**. The results indicated that most Fe_2O_3/TiO_2 films showed a lower emission intensity. Hence, the Fe_2O_3/TiO_2 heterostructures would have a low recombination rate of electrons and holes, improving the photocatalytic activity of the obtained films.

Furthermore, **Figure 4b** presents the degradation profile plotted as C/C_0 versus irradiation time, where C is the concentration of RhB at the irradiation time (t in h), and C_0 represents the initial concentration of RhB. After 5 h of irradiation, rhodamine was degraded to a maximum of 63% using Fe_2O_3/TiO_2 heterostructured photocatalysts. In addition, the stability of films was also studied, wherein after three cycles of use, the efficiency of RhB degradation was decreased by 28% compared to the first usage.

Costa et al. [44] investigated the photocurrent response of heterostructured photoanodes composed of Fe_2O_3/WO_3 and WO_3/Fe_2O_3, deposited onto FTO-glass. The heterostructure films were investigated as photocatalyst material for Rhodamine B (RhB) dye degradation in an aqueous solution. Significantly, the order of the semiconductor layer influences the flat-band potential (E_{fb}) positions and consequently changes the charge mobility in the electrode (photoanode). Under this study's conditions, the heterostructured WO_3/Fe_2O_3 film showed reduced charge recombination and increased photocurrent. Therefore, WO_3/Fe_2O_3 heterostructure film showed superior photoelectrocatalytic efficiency for RhB dye degradation (32%) in comparison to Fe_2O_3 or Fe_2O_3/WO_3 heterostructure film, as shown in **Figure 5**.

Heterostructures based on Fe_2O_3 can also be used to decompose toxic metal ions from wastewater as Cr, Pb, and Hg [45–47]. Particularly, the presence of Cr(VI) above 0.05 mg/L (WHO standard) in drinking water is lethal mutagenic and carcinogenic to human beings [48]. In comparison to Cr (VI), the trivalent chromium (Cr(III)) is less harmful. Therefore, reducing Cr(VI) to Cr(III) has become an emerging research

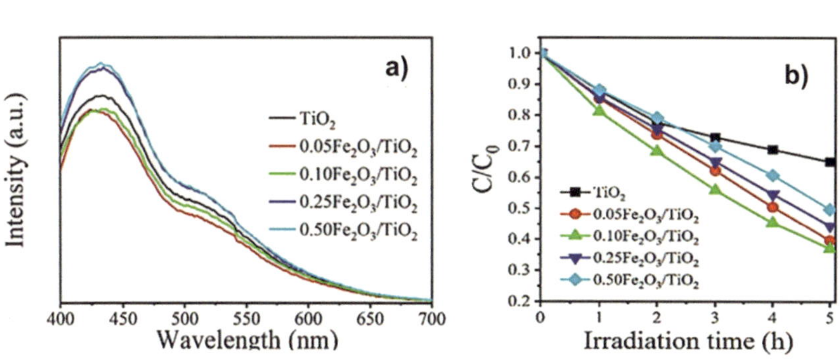

Figure 4.
(a) Photoluminescence (PL) spectra of Fe_2O_3/TiO_2 heterostructures and (b) Photodegradation curves of RhB over Fe_2O_3/TiO_2 heterostructures photocatalysts. Reproduced with permission from [42].

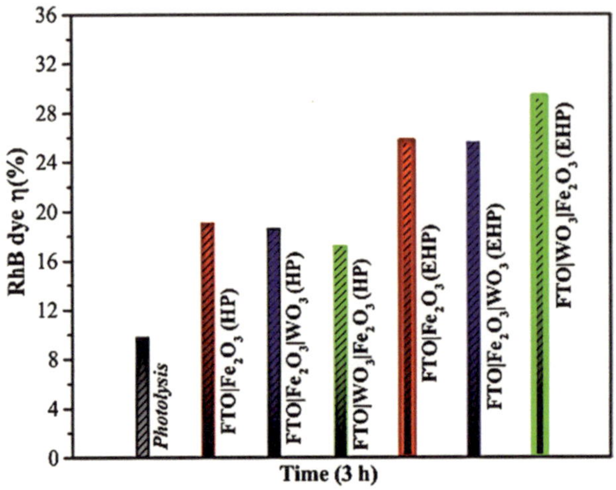

Figure 5.
Efficiency in the degradation of RhB dye in aqueous solution during polychromatic irradiation by photolysis, heterogeneous photocatalysis (HP), and electro-assisted heterogeneous photocatalysis (EHP) using FTO/Fe_2O_3, FTO/Fe_2O_3/WO_3, and FTO/WO_3/Fe_2O_3 electrodes. Reproduced with permission from [44].

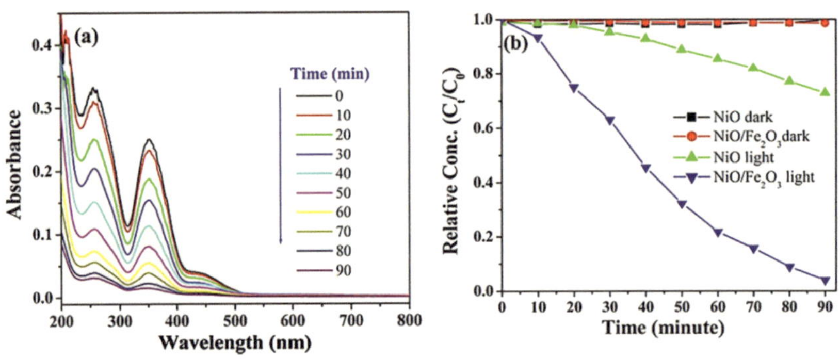

Figure 6.
(a) Photocatalytic reduction of 5 × 10^{-4} M Cr(VI) in the presence of NiO/Fe_2O_3 heterostructure thin films. (b) Reaction profile of Cr(VI) reduction against specific time intervals for the different samples obtained. Reproduced with permission from [49].

topic in the environmental field. Consequently, Jana et al. [49] constructed a novel heterostructure based on NiO/Fe_2O_3 thin film, which was employed in the photoreduction of Cr(VI). Thin films were used in this research to overcome the drawbacks related to material recovery and agglomeration during the photocatalysis process observed when powders are used as photocatalysts [50].

The photocatalytic reduction spectra of aqueous Cr(VI) (5 × 10^{-4} mol/L) at pH 2 under visible light irradiation in the presence of NiO/Fe_2O_3 heterostructures is shown in **Figure 6**. The results reveal that the absorption spectra of Cr(VI) centered at 350 nm decrease with the exposure time (**Figure 6a**), as well as the NiO/Fe_2O_3 thin films show the remarkable photo-reduction ability of approximately 100% within 90 min of visible light irradiation as shown in **Figure 6b**. A blank experiment (i.e., only light irradiation without catalyst) showed that no change in absorption of Cr(VI) was observed under the control experimental conditions, corroborating that the removal of aqueous Cr(VI) was truly driven by a photocatalytic process, rather than simple physical adsorption of Cr(VI).

Photocatalytic reduction of Cr(VI) to Cr (III) is principally promoted by photoexcited electrons generated after light illumination. The proposed schematic diagram can explain the possible mechanism well, as illustrated in **Figure 7**. When the NiO/Fe_2O_3 thin films are irradiated from the light source, photogenerated electrons on the conduction band (CB) of Fe_2O_3 are transferred to the CB of NiO to decrease potential energy, and holes on the valence band (VB) of NiO migrate to the VB of Fe_2O_3 simultaneously. After that, CB electrons of NiO readily react with O_2 forming O_2^- and other active species. Furthermore, CB electrons also directly react with Cr(V) and Cr(IV), in agreement with the observations proposed in previous works [51, 52]. This easy cyclic transport of electron-hole within NiO/Fe_2O_3 matrix enables the separation of photogenerated electrons and holes, which reduces electron-hole recombination, thus improving the photoreduction efficiency of the heterostructure.

Furthermore, the stability of the as-synthesized thin films after photocatalytic reduction of Cr(VI) was studied. The results reveal that the photocatalytic activity of NiO/Fe_2O_3 thin films decreases after six cycles of photoreduction of Cr(VI), attributed mainly due to the deposition of small amounts of Cr(III) onto the surface of NiO/Fe_2O_3 after each cycle.

3. Heterostructures based on CuO

Copper oxide (CuO) is an important p-type semiconductor with a narrow bandgap of 1.3–2.2 eV, large absorption of visible light, low cost, uncomplicated fabrication, and low toxicity. It has been widely investigated as photocatalysts in the photocatalytic degradation of organic pollutants in wastewater [53, 54]. Besides, CuO generates a high photocurrent compared to α-Fe_2O_3 [54]. However, the photocatalytic activity of the bare CuO is still low due to its high photoexcited electron-hole recombination [55]. Therefore, prevention/slowdown of the rate of recombination of photogenerated electron-hole pairs is an important parameter in aiming for superior photocatalytic activity. Recently, creating a heterostructure by coupling an n-type semiconductor

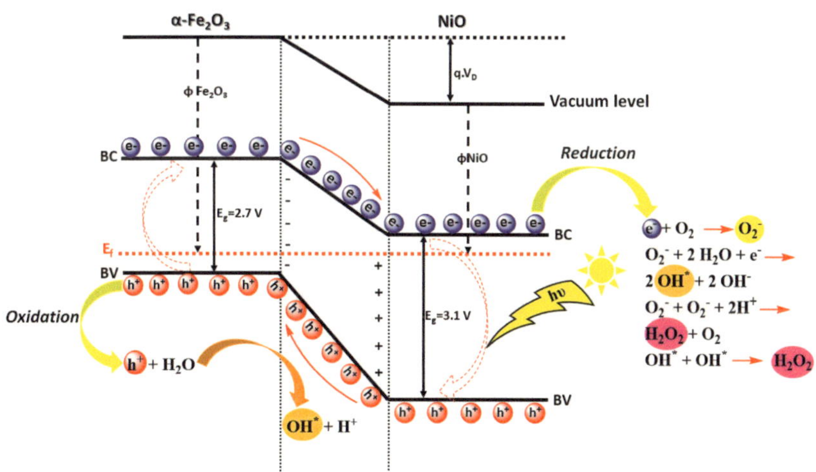

Figure 7.
Schematic representation of band diagram and photodecomposition pathway of NiO/Fe_2O_3 heterostructure photocatalyst; where ϕ, V_D, E_F, and E_g represent work function, contact potential, Fermi level and band gap. Figure adapted from [49].

with p-type CuO may represent an effective strategy for enhancing the stability of the photogenerated electron-hole pairs. Indeed, many studies have been focused on developing heterostructure CuO-based composite photocatalysts for photocatalytic applications, such as CuO/ZnO, CuO/TiO_2, CuO/Cu_2O, CuO/WO_3, CuO/CeO_2 and CuO/SnO_2 [56–61].

For example, Selleswari et al. [61] fabricated CuO/SnO_2 heterostructure thin films by spray pyrolysis technique and found that the efficiency of photodegradation of Congo-red (CR) and Malachite green (MG) under UV light irradiation was improved compared to pristine CuO and SnO_2. The degradation efficiency of the CuO/SnO_2 (1:1 ratio) heterostructure was 90 and 97% for the MG and CR dyes, respectively (**Figure 8a** and **b**, respectively). Besides, after seven photocatalytic degradation cycles, no significant decrease in the photocatalytic activity of the heterostructure was observed (only a loss of 3–5%), as shown in **Figure 8c** and **d**.

The enhanced photocatalytic activity of the heterostructure is attributed to its synergistic action on the specific adsorption property and low recombination probability of photo-generated carriers due to the efficient charge transfer between CuO and SnO_2.

Additionally, ZnO is another semiconductor used as a coupling for CuO-based photocatalysts. In particular, Nguyen et al. [62] investigated the photocatalytic activity of ZnO/CuO thin films fabricated onto a glass substrate by sputtering, thermal

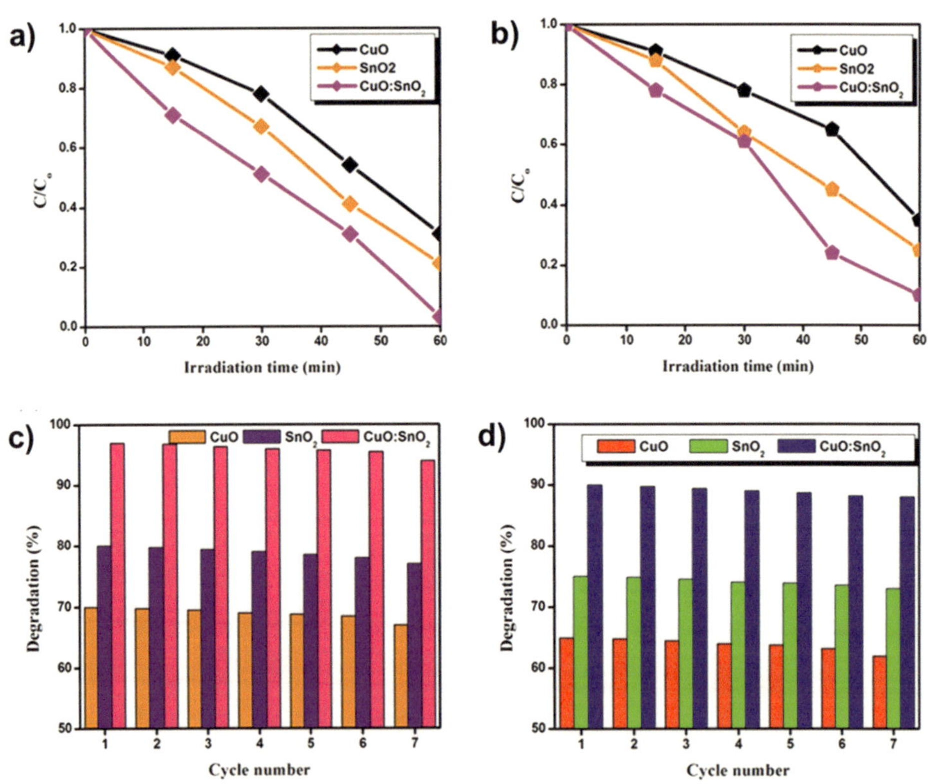

Figure 8.
Degradation profile of (a) Congo-red and (b) Malachite green dyes; seven cycles segment for degradation of (c) Congo-red and (d) Malachite green under UV light irradiation for CuO, SnO_2, and CuO/SnO_2. Reproduced with permission from [61].

annealing, spin coating, and simple hydrothermal methods. The optical absorption analysis shown in the spectra in **Figure 9a** reveals that the ZnO/CuO heterostructure film has a broad absorption range and the highest optical absorption compared to ZnO film and CuO film. This phenomenon can be explained by the attributions of the high surface roughness of the ZnO film in the heterostructured ZnO/CuO film, which can reduce the optical reflection on the surface of the composite film [62]. Furthermore, the photocatalytic activities of the fabricated samples were examined by the rate of degradation of RhB dye under simulated solar irradiation. **Figure 9b** shows the degradation efficiencies calculated for CuO, ZnO, and ZnO/CuO thin films, where the values obtained after 120 min of illumination were 55, 78, and 93%, respectively. The improvement in the degradation efficiency of this heterostructured film was mainly ascribed to the effective suppression of the electron-hole pairs recombination and its higher photon absorption. Besides, **Figure 9c** shows the cycling photodegradation of the ZnO/CuO thin film, obtaining that after three cycling experiments, the film maintains good degradation efficiency for dye contamination, demonstrating that this new material is a highly photostable and reusable photocatalyst.

Another heterostructures based on CuO and CuO_2 thin films used in the degradation of organic pollutants was proposed by Khiavi et al. [58]. In this research, a photocatalyst was developed by integrating cupric oxide (CuO) and cuprous oxide (Cu_2O) thin films, which showed superior performance for the photocatalytic degradation of methylene blue (MB) compared to CuO and Cu_2O pristine photocatalysts. As shown in **Figure 10a**, Cu_2O was deposited on top of the CuO thin films due to the lower optical absorption of Cu_2O than CuO, a greater bandgap of Cu_2O (~2.2 eV) than CuO (~1.6 eV), and a longer carrier diffusion length of Cu_2O (~500 nm) than CuO (~200 nm). The MB concentration change as a function of the photocatalytic

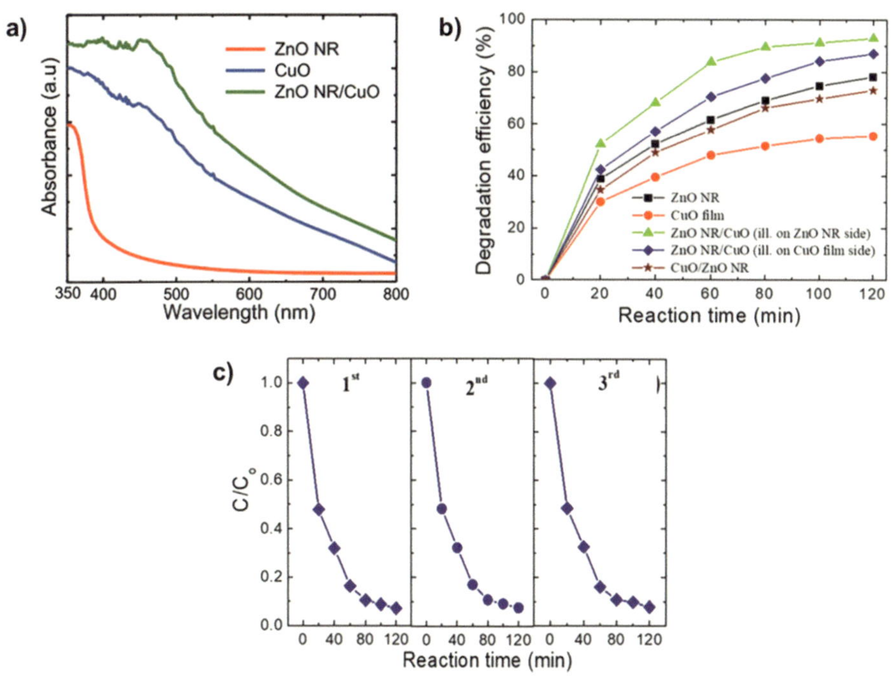

Figure 9.
(a) Absorption spectra; (b) photodegradation of RhB; (c) recycling photodegradation of CuO, ZnO film, and ZnO/CuO heterostructure thin films. Reproduced with permission from [62].

degradation time under visible light was plotted and is shown in **Figure 10b**. The results showed that the MB degradation rate by the CuO/Cu_2O heterostructure was almost twice that of the bare semiconductors. The enhanced photocatalytic activity for CuO/Cu_2O heterostructures can be attributed to the improved separation of photogenerated electrons and holes, where the interface between Cu_2O and CuO acted as a key parameter for this separation. Meanwhile, higher absorption of visible light was achieved based on the difference in the energy levels of their conduction bands and valence bands.

Photocatalytic reduction for removing toxic metals from wastewater is considered one of the most attractive methods because of its low cost and ease of operation [63–65]. In particular, heterostructures based on CuO are suggested as promising materials for metal ion reduction under visible illumination [66–68]. For instance, Ghosh and Mondal [69] developed a unique binary heterostructure-based photocatalyst consisting of Cu_7S_4/CuO and explored its application in the photocatalytic reduction of Ni (II) under visible light. The results observed in **Figure 11a** reveal that Cu_7S_4/CuO films exhibit better photocatalytic performance than pure Cu_7S_4 and CuO films. The removal rate of Ni(II) for pure Cu_7S_4 film is 84% at 90 min and 89% for pure CuO films at 75 min, while for the Cu_7S_4/CuO heterostructure, the removal rate increased to 95% at 60 min. Likewise, these authors fabricated a thin-film TiO_2/CuO

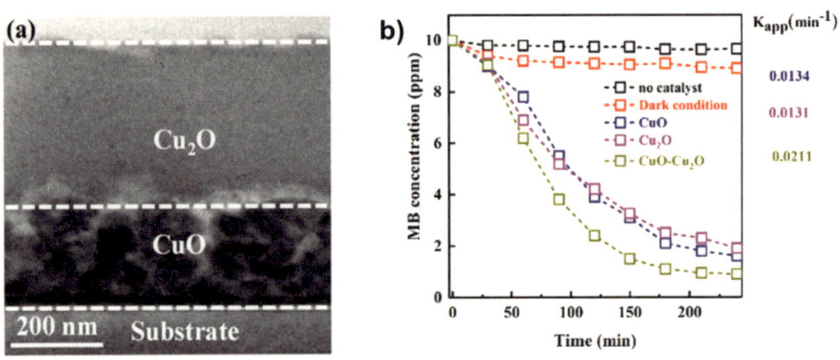

Figure 10.
(a) Cross-sectional TEM image of the fabricated heterostructure CuO/Cu_2O thin film photocatalyst; (b) MB degradation profiles for different irradiation times for CuO, Cu_2O, and CuO/Cu_2O samples. Reproduced with permission from [58].

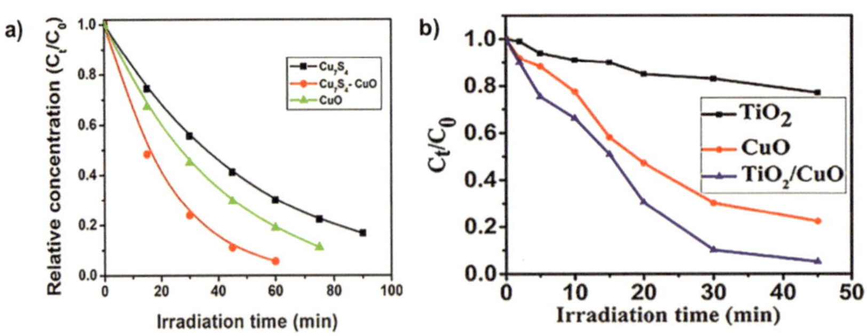

Figure 11.
Reaction profile for the photocatalytic reduction of Ni (II) in the presence of (a) Cu_7S_4/CuO and (b) TiO_2/CuO heterostructure films under visible light irradiation. Reproduced with permission from [69, 70].

heterostructure on an FTO glass substrate [70] and found a faster reduction of Ni(II) than in single films. The total decrease of Ni^{2+} by pure TiO_2 was only 23% after 45 min of light irradiation, and for pure CuO, it extended up to 78% for the same period (**Figure 11b**). Meanwhile, for TiO_2/CuO heterostructure, the degradation further increased up to 95% simultaneously.

The increase in photocatalytic performance for heterostructures in both cases happened due to the effective separation of electron-hole pairs by the junction to lower their recombination rate, thus, paving the way for better reduction. Specifically, a possible mechanism of Ni^{2+} reduction by the Cu_7S_4/CuO heterostructure catalyst was described. When both semiconductors are simultaneously excited by the absorption of photons under visible light irradiation, electron-hole pairs are created. Subsequently, the photogenerated electrons from the conduction band of Cu_7S_4 get easily transferred to the conduction band of CuO. While, the respective holes in the valence band of CuO would move upward in energy to the VB of Cu_7S_4, which retards the recombination of photogenerated electron-hole pairs. Then, the reduction of Ni(II) ions to Ni(0) can occur by the photogenerated electrons on the heterostructure surface. At the same time, the holes oxidize water to oxygen or organic compounds to CO_2.

Moreover, Pastrana et al. [71] fabricated heterostructures based on CuO/α-Fe_2O_3 by dip-coating technique. Due to their rough structure, these heterostructures present efficient and fast arsenic removal performances compared to pure oxides. The removal of arsenic was attributed to the direct absorption of As(III) on thin films and the photocatalytic oxidation of As(III) to As (V). The results shown in **Figure 12a** indicate that the highest absorption efficiency obtained for the CuO/Fe_2O_3 heterostructures was approximately 85% within the first 20 min of irradiation; after that, the arsenite removal efficiency remains relatively constant. Besides, removal tests in the darkness were performed, obtaining removal efficiency below 10% (see **Figure 12b**), corroborating that the heterostructures photoactivity effectively enhanced the adsorption. The higher removal efficiency of heterostructures compared to pristine oxides is mainly due to roughness and the slower recombination of electron-hole pairs for the heterostructures. Many researchers claim that the increase of the surface roughness may favor the adsorption of molecules on the surface films [71] and that the inhibition of recombination of charge carriers will produce a more significant amount of reactive oxygen species, which are intermediates to oxidize As(III) to As(V) [72].

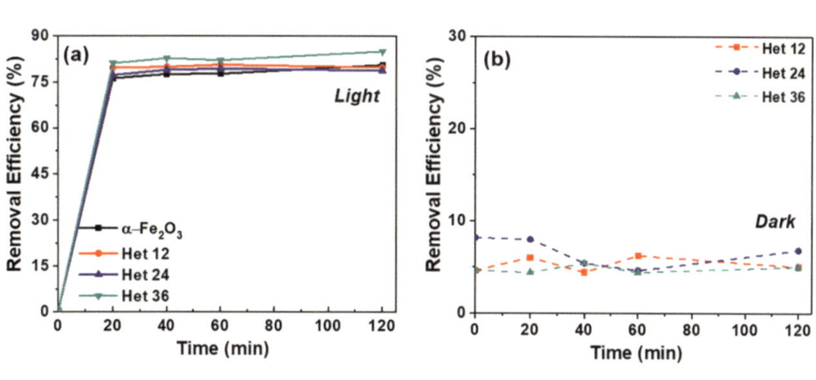

Figure 12.
Effect of contact time variation on As(III) removal efficiency on α-Fe_2O_3 and α-Fe_2O_3/CuO heterostructure thin films. Reproduced with permission from [71].

4. Summary and outlook for future directions

In this review, we presented an overview of research developments in the fabrication of heterostructures based on CuO and α-Fe_2O_3 thin films for photocatalytic applications in wastewater treatment, such as photodegradation of organic pollutants and photoreduction of metal ions. According to the present study, a great number of researches related to efficient heterostructured photocatalysts based on CuO and Fe_2O_3 thin films have been reported. The improvement in photocatalytic activity is mainly attributed to the better light absorption and the efficient separation/transport of electron-hole pairs, and some examples are highlighted in this review. However, even though significant progress has been achieved in studying CuO-based and Fe_2O_3-based heterostructured photocatalysts based, improving these thin film-type heterostructured photocatalysts still presents challenges. Therefore, to resolve these drawbacks, it is necessary to consider the following aspects carefully: (1) the quantum efficiency of these heterostructures is still low compared to utilizing solar energy with high efficiency. Thus, a better understanding of the dynamic behavior of photogenerated carriers in the interface and surface of these types of heterostructures is required; (2) the catalyst's reusability, longevity, and stability should be emphasized in terms of practical applications and (3) more research should be developed to improve the scale production of these heterostructures, as well as its economic feasibility and long-term durability for the treatment of real industrial wastewater in the coming years.

In addition, we also propose some areas for future investigation, for instance, (1) to overcome disadvantages and improve their catalytic performance, the heterostructures based on CuO and Fe_2O_3 thin films can be modified via doping and the formation of ternary heterostructures by combining them with other semiconductors, graphene or other carbon-based materials; (2) the production of these heterostructures onto a variety of substrates, could be another interesting approach for photoelectrocatalysis or photocatalysis applications, as well as convenient to reuse the catalyst; (3) guide future research in the use natural sunlight, due to its abundance, as a source of illumination instead of commercially simulated sunlight and (4) carrying out research related to computational calculation and design forms of heterostructures based on CuO and Fe_2O_3 thin films will us get an extensive perception of the system and charge transportation kinetics in these materials, as well as clarify the correct vision of photocatalytic processes in CuO-based and Fe_2O_3-based heterostructured photocatalysts thin films.

In conclusion, this article has summarized the current efficient developments in the fabrication and application of photocatalysts of CuO-based and Fe_2O_3-based heterostructures thin films. Both CuO and Fe_2O_3 are non-toxic, cost-effective, and photoactive materials under UV/visible/solar irradiation, making them recommended semiconductors for heterostructures formation. However, it is still necessary to improve certain drawbacks in search of the more remarkable development of heterostructured systems based on CuO and Fe_2O_3 thin films with high efficiency and stability. If these objectives are achieved, these heterostructures will be used in industrial water treatment systems in some years. Finally, the implicit final aim of this review was not only to summarize the state of the art but also to stimulate the readers and arouse their curiosity, leading them to investigate this particular class of nanomaterials in greater detail.

Acknowledgements

The work described in this review was financially supported by the projects CONCYTEC under the contract numbers N° 032-2019-FONDECYT-BM-INC. INV, N° 120-2018-FONDECYT, N°08-FONDECYT-BM-IADT-MU-2018, N° 237-2015-FONDECYT and N° 059-2021-PROCIENCIA.

Conflict of interest

The authors declare no conflict of interest.

Author details

Elizabeth C. Pastrana*, Pierre G. Ramos*, Luis A. Sánchez and Juan M. Rodriguez
Center for the Development of Advanced Materials and Nanotechnology, National University of Engineering, Lima, Peru

*Address all correspondence to: epastrana@uni.edu.pe and pramosa@uni.pe

References

[1] Hasija V, Kumar A, Sudhaik A, Raizada P, Singh P, Le QV, et al. Step-scheme heterojunction photocatalysts for solar energy, water splitting, CO_2 conversion, and bacterial inactivation: A review. Environmental Chemistry Letters. 2021;**19**:2941-2966. DOI: 10.1007/s10311-021-01231-w

[2] Barreca D, Comini E, Ferrucci AP, Gasparotto A, Maccato C, Maragno C, et al. First example of $ZnO-TiO_2$ nanocomposites by chemical vapor deposition: Structure, morphology, composition, and gas sensing performances. Chemistry of Materials. 2007;**19**:5642-5649. DOI: 10.1021/cm701990f

[3] Moon WJ, Yu JH, Choi GM. The CO and H_2 gas selectivity of CuO-doped SnO_2–ZnO composite gas sensor. Sensors and Actuators B: Chemical. 2002;**87**:464-470. DOI: 10.1016/S0925-4005(02)00299-X

[4] Vattikuti SVP. Heterostructured nanomaterials: Latest trends in formation of inorganic heterostructures. In: Bhagyaraj SM, Oluwafemi SO, Kalarikkal N, Thomas S, editors. Synthesis of Inorganic Nanomaterials: Advances and Key Technologies. 1st ed. Amsterdam: Elsevier; 2018. pp. 89-120. DOI: 10.1016/B978-0-08-101975-7.00004-X

[5] Roul B, Chandan G, Mukundan S, Krupanidhi SB. Heterostructures of III-nitride semiconductors for optical and electronic applications. In: Zhong M, editor. Epitaxy. 1st ed. London: IntechOpen; 2017. pp. 171-201. DOI: 10.5772/intechopen.70219

[6] Smets AHM, Jäger K, Isabella O, Zeman M. Solar Energy: The Physics and Engineering of Photovoltaic Conversion, Technologies and Systems. Cambridge: UIT; 2016. p. 484

[7] Neamen DA. Semiconductor Physics and Devices: Basic Principles. 3rd ed. New York: McGraw-Hill; 2003. p. 269

[8] Wang H, Zhang L, Chen Z, Hu J, Li S, Wang Z, et al. Semiconductor heterojunction photocatalysts: Design, construction, and photocatalytic performances. Chemical Society Reviews. 2014;**43**:5234-5244. DOI: 10.1039/C4CS00126E

[9] Bhagyaraj SM, Oluwafemi SO, Kalarikkal N, Thomas S, editors. Synthesis of Inorganic Nanomaterials: Advances and Key Technologies. 1st ed. Amsterdam: Elsevier; 2018. p. 300. DOI: 10.1016/C2016-0-01718-7

[10] Zhou J, Tang B, Lin J, Lv D, Shi J, Sun L, et al. Morphology engineering in monolayer MoS_2-WS_2 lateral heterostructures. Advanced Functional Materials. 2018;**28**:1801568. DOI: 10.1002/adfm.201801568

[11] Šutka A, Järvekülg M, Gross KA. Photocatalytic nanoheterostructures and chemically bonded junctions made by solution-based approaches. Critical Reviews in Solid State and Materials Sciences. 2019;**44**:239-263. DOI: 10.1080/10408436.2018.1485549

[12] Zhao X, Li Q, Xu L, Zhang Z, Kang Z, Liao Q, et al. Interface engineering in 1D ZnO-based heterostructures for photoelectrical devices. Advanced Functional Materials. 2021;**32**:2106887. DOI: 10.1002/adfm.202106887

[13] Liu X, Hersam MC. Interface characterization and control of 2D

materials and heterostructures. Advanced Materials. 2018;**30**:1801586. DOI: 10.1002/adma.201801586

[14] Emeline AV, Rudakova AV, Mikhaylov RV, Bulanin KM, Bahnemann DW. Photoactive heterostructures: How they are made and explored. Catalysts. 2021;**11**:294. DOI: 10.3390/catal11020294

[15] Cantele G, Ninno D. Size-dependent structural and electronic properties of Bi (111) ultrathin nanofilms from first principles. Physical Review Materials. 2017;**1**:014002. DOI: 10.1103/PhysRevMaterials.1.014002

[16] Fast J, Barrigon E, Kumar M, Chen Y, Samuelson L, Borgström M, et al. Hot-carrier separation in heterostructure nanowires observed by electron-beam induced current. Nanotechnology. 2020;**31**:394004. DOI: 10.1088/1361-6528/ab9bd7

[17] Guari Y, Cahu M, Félix G, Sene S, Long J, Chopineau J, et al. Nanoheterostructures based on nanosized Prussian blue and its analogues: Design, properties and applications. Coordination Chemistry Reviews. 2022;**461**:214497. DOI: 10.1016/j.ccr.2022.214497

[18] Wang X, Tian W, Liao M, Bando Y, Golberg D. Recent advances in solution-processed inorganic nanofilm photodetectors. Chemical Society Reviews. 2014;**43**:1400-1422. DOI: 10.1039/C3CS60348B

[19] Wang L, Shi C, Wang L, Pan L, Zhang X, Zou JJ. Rational design, synthesis, adsorption principles and applications of metal oxide adsorbents: A review. Nanoscale. 2020;**12**:4790-4815. DOI: 10.1039/C9NR09274A

[20] Tadic M, Panjan M, Tadic BV, Kralj S, Lazovic J. Magnetic properties of mesoporous hematite/alumina nanocomposite and evaluation for biomedical applications. Ceramics International. 2022;**48**:10004-10014. DOI: 10.1016/j.ceramint.2021.12.209

[21] Li C, Luo Z, Wang T, Gong J. Surface, bulk, and interface: Rational design of hematite architecture toward efficient photo-electrochemical water splitting. Advanced Materials. 2018;**30**:1707502. DOI: 10.1002/adma.201707502

[22] Shen S, Lindley SA, Chen X, Zhang JZ. Hematite heterostructures for photoelectrochemical water splitting: Rational materials design and charge carrier dynamics. Energy & Environmental Science. 2016;**9**:2744-2775. DOI: 10.1039/C6EE01845A

[23] Salari H, Sadeghinia M. MOF-templated synthesis of nano Ag_2O/ZnO/CuO heterostructure for photocatalysis. Journal of Photochemistry and Photobiology A: Chemistry. 2019;**376**:279-287. DOI: 10.1016/j.jphotochem.2019.03.010

[24] Zhang Q, Zhang K, Xu D, Yang G, Huang H, Nie F, et al. CuO nanostructures: Synthesis, characterization, growth mechanisms, fundamental properties, and applications. Progress in Materials Science. 2014;**60**:208-337. DOI: 10.1016/j.pmatsci.2013.09.003

[25] Spencer JA, Mock AL, Jacobs AG, Schubert M, Zhang Y, Tadjer MJ. A review of band structure and material properties of transparent conducting and semiconducting oxides: Ga_2O_3, Al_2O_3, In_2O_3, ZnO, SnO_2, CdO, NiO, CuO, and Sc_2O_3. Applied Physics Reviews. 2022;**9**:011315. DOI: 10.1063/5.0078037

[26] Yu L, Zhang Y, He J, Zhu H, Zhou X, Li M, et al. Enhanced photoelectrochemical properties of

a-Fe_2O_3 nanoarrays for water splitting. Journal of Alloys and Compounds. 2018;**753**:601-606. DOI: 10.1016/j.jallcom.2018.04.258

[27] Mirzaei A, Hashemi B, Janghorban K. α-Fe_2O_3 based nanomaterials as gas sensors. Journal of Materials Science: Materials in Electronics. 2016;**27**:3109-3144. DOI: 10.1007/s10854-015-4200-z

[28] Tamirat AG, Rick J, Dubale AA, Su WN, Hwang BJ. Using hematite for photoelectrochemical water splitting: A review of current progress and challenges. Nanoscale Horizons. 2016;**1**:243-267. DOI: 10.1039/C5NH00098J

[29] Akter J, Hanif MA, Islam MA, Sapkota KP, Lee I, Hahn JR. Visible-light-active novel α-Fe_2O_3/Ta_3N_5 photocatalyst designed by band-edge tuning and interfacial charge transfer for effective treatment of hazardous pollutants. Journal of Environmental Chemical Engineering. 2021;**9**:106831. DOI: 10.1016/j.jece.2021.106831

[30] Zheng Y, Zhang X, Zhao J, Yang P. Assembled fabrication of α-Fe_2O_3/BiOCl heterojunctions with enhanced photocatalytic performance. Applied Surface Science. 2018;**430**:585-594. DOI: 10.1016/j.apsusc.2017.06.097

[31] Mishra M, Chun DM. α-Fe_2O_3 as a photocatalytic material: A review. Applied Catalysis A: General. 2015;**498**:126-141. DOI: 10.1016/j.apcata.2015.03.023

[32] Cheng L, Liu L, Li R, Zhang J. Liquid phase deposition of α-Fe_2O_3/ZnO heterojunction film with enhanced visible-light photoelectrocatalytic activity for pollutant removal. Journal of The Electrochemical Society. 2015;**498**:126-141. DOI: 10.1016/j.apcata.2015.03.023

[33] Jinhui J, Kuili L, Weiqiang F, Meng L, Yu L, Baodong M, et al. Electrospinning synthesis and photocatalytic property of Fe_2O_3/$MgFe_2O_4$ heterostructure for photocatalytic degradation of tetracycline. Materials Letters. 2016;**176**:1-4. DOI: 10.1016/j.matlet.2016.03.016

[34] Tao Q, Bi J, Huang X, Wei R, Wang T, Zhou Y, et al. Fabrication, application, optimization and working mechanism of Fe_2O_3 and its composites for contaminants elimination from wastewater. Chemosphere. 2021;**263**:127889. DOI: 10.1016/j.chemosphere.2020.127889

[35] Yu Z, Chouchene B, Liu M, Moussa H, Schneider R, Moliere M, et al. Influence of laminated architectures of heterostructured CeO_2-ZnO and Fe_2O_3-ZnO films on photodegradation performances. Surface and Coatings Technology. 2020;**403**:126367. DOI: 10.1016/j.surfcoat.2020.126367

[36] Gao M, Zhu L, Ong WL, Wang J, Ho GW. Structural design of TiO_2-based photocatalyst for H_2 production and degradation applications. Catalysis Science & Technology. 2015;**5**:4703-4726. DOI: 10.1039/C5CY00879D

[37] Suryavanshi RD, Mohite SV, Bagade AA, Rajpure KY. Photoelectrocatalytic activity of spray deposited Fe_2O_3/ZnO photoelectrode for degradation of salicylic acid and methyl orange dye under solar radiation. Materials Science and Engineering: B. 2019;**248**:114386. DOI: 10.1016/j.mseb.2019.114386

[38] Suryavanshi RD, Rajpure KY. Spray deposited Fe_2O_3 and stratified Fe_2O_3/ZnO novel photoelectrode for photoelectrocatalytic degradation of benzoic acid under solar light illumination. Journal of Photochemistry

and Photobiology A: Chemistry. 2018;**357**:72-80. DOI: 10.1016/j.jphotochem.2018.02.008

[39] Yilmaz C, Unal U. Morphology and crystal structure control of α-Fe_2O_3 films by hydrothermal-electrochemical deposition in the presence of Ce^{3+} and/or acetate, F^- ions. RSC Advances. 2016;**6**:8517-8527. DOI: 10.1039/C5RA20105E

[40] Mandal S, Adhikari S, Pu S, Wang X, Kim DH, Patel RK. Interactive Fe_2O_3/porous SiO_2 nanospheres for photocatalytic degradation of organic pollutants: Kinetic and mechanistic approach. Chemosphere. 2019;**234**:596-607. DOI: 10.1016/j.chemosphere.2019.06.092

[41] Ramos PG, Sánchez LA, Rodriguez JM. A review on improving the efficiency of photocatalytic water decontamination using ZnO nanorods. Journal of Sol-Gel Science and Technology. 2022;**102**:105-124. DOI: 10.1007/s10971-021-05707-7

[42] Wannapop S, Somdee A, Bovornratanaraks T. Experimental study of thin film Fe_2O_3/TiO_2 for photocatalytic Rhodamine B degradation. Inorganic Chemistry Communications. 2021;**128**:108585. DOI: 10.1016/j.inoche.2021.108585

[43] Moniz SJA, Shevlin SA, Martin DJ, Guo ZX, Tang J. Visible-light driven heterojunction photocatalysts for water splitting—A critical review. Energy & Environmental Science. 2015;**8**:731-759. DOI: 10.1039/C4EE03271C

[44] Costa GS, Costa MJS, Oliveira HG, Lima LCB, JrGE L, Cavalcante LS, et al. Effect of the applied potential condition on the photocatalytic properties of Fe_2O_3|WO_3 heterojunction films. Journal of Inorganic and Organometallic Polymers and Materials. 2020;**30**:2851-2862. DOI: 10.1007/s10904-019-01429-0

[45] Li K, Liang Y, Yang J, Zhang H, Yang G, Lei W. BiOCl/Fe_2O_3 heterojunction nanoplates with enhanced visible-light-driven photocatalytic performance for degrading organic pollutants and reducing Cr(VI). Journal of Photochemistry and Photobiology A: Chemistry. 2018;**364**:240-249. DOI: 10.1016/j.jphotochem.2018.06.001

[46] Wei J, Zhao J, Li CY, Xie XY, Wei YY, Shen W, et al. Highly sensitive and selective electrochemical detection of Pb(II) in serum via an α-Fe_2O_3/NiO heterostructure: Evidence from theoretical calculations and adsorption investigation. Sensors and Actuators B: Chemical. 2021;**344**:130295. DOI: 10.1016/j.snb.2021.130295

[47] Kadi MW, Mohamed RM, Ismail AA, Bahnemann DW. Performance of mesoporous α-Fe_2O_3/g-C_3N_4 heterojunction for photoreduction of Hg(II) under visible light illumination. Ceramics International. 2020;**46**:23098-23106. DOI: 10.1016/j.ceramint.2020.06.087

[48] Sharma P, Singh SP, Parakh SK, Tong YW. Health hazards of hexavalent chromium (Cr (VI)) and its microbial reduction. Bioengineered. 2022;**13**:4923-4938. DOI: 10.1016/j.ceramint.2020.06.087

[49] Jana S, Mondal A, Ghosh A. Fabrication of stable NiO/Fe_2O_3 heterostructure: A versatile hybrid material for electrochemical sensing of glucose, methanol and enhanced photodecomposition and/photoreduction of water contaminants. Applied Catalysis B: Environmental. 2018;**232**:26-36. DOI: 10.1016/j.apcatb.2018.03.038

[50] Pedanekar RS, Shaikh SK, Rajpure KY. Thin film photocatalysis for environmental remediation: A status review. Current Applied Physics. 2020;**20**:931-952. DOI: 10.1016/j.cap.2020.04.006

[51] Zhao Z, An H, Lin J, Feng M, Murugadoss V, Ding T, et al. Progress on the photocatalytic reduction removal of chromium contamination. The Chemical Record. 2018;**19**:873-882. DOI: 10.1002/tcr.201800153

[52] Alharbi NS, Hu B, Hayat T, Rabah SO, Alsaedi A, Zhuang L, et al. Efficient elimination of environmental pollutants through sorption-reduction and photocatalytic degradation using nanomaterials. Frontiers of Chemical Science and Engineering. 2020;**14**:1124-1135. DOI: 10.1007/s11705-020-1923-z

[53] Malwal D, Gopinath P. Enhanced photocatalytic activity of hierarchical three dimensional metal oxide@CuO nanostructures towards the degradation of Congo red dye under solar radiation. Catalysis Science & Technology. 2016;**6**:4458-4472. DOI: 10.1039/C6CY00128A

[54] Pastrana EC, Zamora V, Wang D, Alarcón H. Fabrication and characterization of α-Fe_2O_3/CuO heterostructure thin films via dip-coating technique for improved photoelectrochemical performance. Advances in Natural Sciences: Nanoscience and Nanotechnology. 2019;**10**:035012. DOI: 10.1088/2043-6254/ab3d2f

[55] Choi J, Oh H, Han SW, Ahn S, Noh J, Park JB. Preparation and characterization of graphene oxide supported Cu, Cu_2O, and CuO nanocomposites and their high photocatalytic activity for organic dye molecule. Current Applied Physics. 2017;**17**:137-145. DOI: 10.1016/j.cap.2016.11.020

[56] Zhu L, Li H, Liu Z, Xia P, Xie Y, Xiong D. Synthesis of the 0D/3D CuO/ZnO Heterojunction with Enhanced Photocatalytic Activity. The Journal of Physical Chemistry C. 2018;**122**:9531-9539. DOI: 10.1021/acs.jpcc.8b01933

[57] Ren S, Zhang J, Lu H, Gao S, Li L, Rong P, et al. 3D Carambola-Like CuO/TiO_2 nanotube heterostructures via low-temperature solution process for photocatalytic activity. European Journal of Inorganic Chemistry. 2021;**2021**:1852-1857. DOI: 10.1002/ejic.202100110

[58] Khiavi ND, Katal R, Eshkalak SK, Masudy-Panah S, Ramakrishna S, Jiangyong H. Visible light driven heterojunction photocatalyst of CuO–Cu_2O thin films for photocatalytic degradation of organic pollutants. Nanomaterials. 2019;**9**:1011. DOI: 10.3390/nano9071011

[59] Dursun S, Koyuncu SN, Kaya IC, Kaya GG, Kalem V, Akyildiz H. Production of CuO–WO_3 hybrids and their dye removal capacity/performance from wastewater by adsorption/photocatalysis. Journal of Water Process Engineering. 2020;**36**:101390. DOI: 10.1016/j.jwpe.2020.101390

[60] Li Z, Liu D, Huang W, Sun Y, Li S, Wei X. Applying facilely synthesized CuO/CeO_2 photocatalyst to accelerate methylene blue degradation in hypersaline wastewater. Surface and Interface Analysis. 2019;**51**:336-344. DOI: 10.1002/sia.6585

[61] Selleswari D, Meena P, Mangalaraj D. Design of CuO/SnO_2 heterojunction photocatalyst with enhanced UV light-driven photocatalytic activity on congo-red and malachite green dyes. Journal of the Iranian Chemical Society. 2019;**16**:1291-1300. DOI: 10.1007/s13738-019-01606-5

[62] Nguyen DT, Tran MD, Hoang TV, Trinh DT, Pham DT, Nguyen DL. Experimental and numerical study on photocatalytic activity of the ZnO nanorods/CuO composite film. Scientific Reports. 2020;**10**:7792. DOI: 10.1038/s41598-020-64784-w

[63] Wahyuni ET, Aprilita NH, Hatimah H, Wulandari AM, Mudasir M. Removal of toxic metal ions in water by photocatalytic method. American Chemical Science Journal. 2015;**5**:194-201. DOI: 10.9734/ACSj/2015/13807

[64] Litter MI. Last advances on TiO_2-photocatalytic removal of chromium, uranium and arsenic. Current Opinion in Green and Sustainable Chemistry. 2017;**6**:150-158. DOI: 10.1016/j.cogsc.2017.04.002

[65] Shrestha R, Ban S, Devkota S, Sharma S, Joshi R, Tiwari AP, et al. Technological trends in heavy metals removal from industrial wastewater: A review. Journal of Environmental Chemical Engineering. 2021;**9**:105688. DOI: 10.1016/j.jece.2021.105688

[66] Mohamed RM, Ismail AA. Photocatalytic reduction and removal of mercury ions over mesoporous CuO/ZnO S-scheme heterojunction photocatalyst. Ceramics International. 2021;**47**:9659-9667. DOI: 10.1016/j.ceramint.2020.12.105

[67] Nogueira AE, Lopes OF, Neto ABS, Ribeiro C. Enhanced Cr(VI) photoreduction in aqueous solution using Nb_2O_5/CuO heterostructures under UV and visible irradiation. Chemical Engineering Journal. 2017;**312**:220-227. DOI: 10.1016/j.cej.2016.11.135

[68] Kadi MW, Mohamed RM, Ismail AA, Bahnemann DW. Soft and hard templates assisted synthesis mesoporous CuO/g-C_3N_4 heterostructures for highly enhanced and accelerated Hg(II) photoreduction under visible light. Journal of Colloid and Interface Science. 2020;**580**:223-233. DOI: 10.1016/j.jcis.2020.07.001

[69] Ghosh A, Mondal A. Efficient charge separation in mixed phase Cu_7S_4-CuO thin film: Enhanced photocatalytic reduction of aqueous Ni (II) under visible-light. Thin Solid Films. 2017;**628**:68-74. DOI: 10.1016/j.tsf.2017.03.001

[70] Ahamed ST, Ghosh A, Show B, Mondal A. Fabrication of n-TiO_2/p-CuO thin-film heterojunction for efficient photocatalytic degradation of toxic organic dyes and reduction of metal ions in solution. Journal of Materials Science: Materials in Electronics. 2020;**31**:16616-16633. DOI: 10.1007/s10854-020-04217-6

[71] Pastrana EC, Loarte SJ, Gonzales-Lorenzo CD, Alta RYP, Alarcón HA. Fabrication and characterization of copper (II) oxide/iron (III) oxide thin film heterostructures for trace arsenic (III) removal in water. Thin Solid Films. 2021;**717**:138440. DOI: 10.1016/j.tsf.2020.138440

[72] Sun T, Zhao Z, Liang Z, Liu J, Shi W, Cui F. Efficient degradation of p-arsanilic acid with arsenic adsorption by magnetic CuO-Fe_3O_4 nanoparticles under visible light irradiation. Chemical Engineering Journal. 2018;**334**:1527-1536. DOI: 10.1016/j.cej.2017.11.052

Chapter 8

Pulsed Laser Deposition of Transparent Conductive Oxides on UV-NIL Patterned Substrates for Optoelectronic Applications

Marcela Socol, Nicoleta Preda, Carmen Breazu and Oana Rasoga

Abstract

Transparent conductive oxide (TCO) electrodes are key components in the fabrication of optoelectronic devices such as organic photovoltaic cells (OPVs) or organic emitting devices (OLEDs). Pulsed laser deposition (PLD) results in TCO coatings with adequate optical and electrical properties, the preservation of the target chemical composition in the transferred films being the major advantage of this technique. Furthermore, the performance of the optoelectronic devices can be enhanced by patterning the TCO electrodes. Indium tin oxide (ITO) remains the most popular TCO due to its high conductivity and transparency. The scarcity of the indium resources encouraged the efforts to find an alternative to ITO, a promising candidate being Al-doped ZnO (AZO). Therefore, this chapter is focused on PLD deposition of TCO films (ITO and AZO) on patterned glass substrates prepared by ultraviolet nanoimprint lithography (UV-NIL) for obtaining transparent electrodes with improved characteristics, which further can be integrated in optoelectronic applications.

Keywords: pulsed laser deposition, patterned transparent conductive oxides, ITO, AZO, ultraviolet nanoimprint lithography

1. Introduction

Organic optoelectronic device such as organic photovoltaics (OPVs) and organic light-emitting devices (OLEDs) focused over the past few decades the attention of both academia and industries due to the possibility to fabricate flexible, transparent devices on large area using low-cost solution processes, leading to cost-effective production [1, 2]. At this stage, in the OPV field, a major concern regards the fabrication of flexible structures with high efficiencies for various applications [3]. Although, OPV with efficiency over 18% has been reported in 2021 [4], further improvements are still needed for making them a real alternative to other photovoltaic cell (PV) technologies (PV based on silicon, PV based on perovskites, etc.). The improvements can be linked to: (i) the type of the organic materials used in the fabrications of the

PV structures; (ii) the deposition techniques used to obtain the organic component as films; and (iii) the different approaches used for enhancing the absorption in the PV structure such as antireflection coatings, back-reflectors, or the surfaces patterning (texturing) [5, 6]. In the PV structures, the thickness of the organic active film is limited by the low carrier mobility and the short exciton diffusion length [7]. An increase in the film thickness leads to a lowering in the device efficiency, while a decrease in the film thickness results in a poor absorption. Lately, some studies reported that the nanopatterning of the transparent electrodes increases the optical path length of light inside the active material improving the performances of the devices [6, 8].

Different optical approaches and structures such as microlens, nanostructured electrodes, scattering layers were used in the field of OLEDs to improve the light extraction efficiency of the devices [9, 10]. The light extraction efficiency is one of the most important parameters of OLED, defined as the ratio of the total number of photons emitted by the OLED and the total number of photons generated within the organic emitter [10, 11]. Thus, the majority of the generated photons in the organic layers are confined inside the device due to the total internal reflection, which takes place at the glass/air and organic/layer substrate interfaces owing to the mismatch of the refractive index [12]. In this way, almost 30% of the emitted photons are trapped in the glass substrate (glass mode), while a 50% are trapped at the organic/anode interface (waveguide mode). Therefore, various methods were used to extract more efficiently the light from the OLEDs [9, 13].

Transparent conductive electrodes (TCE) play a key role in the development of optoelectronic devices such as OPVs, OLEDs, touch screens, electrochromic devices, heat mirrors, smart windows, and so on [14–16]. Over time, various materials such as metal oxides, ultrathin metals, metal nanowires, graphene, carbon nanotubes, conductive polymers, etc., were deposited and investigated as TCE [1, 14]. However, indium tin oxide (ITO) remains the most commonly used TCE due to its remarkable properties such as high transparency (90% at 550 nm wavelength), adequate sheet resistance (10–30 Ω/□), work function (4.7 eV), and reduced roughness (<1 nm) [17, 18]. Besides that, aluminum-doped zinc oxide (AZO) is a suitable metal oxide for replacing ITO since this material met the necessary criteria regarding the high transparency and the electrical resistivity [19, 20].

Transparent conductive oxide (TCO) films can be deposited by numerous chemical and physical methods such as sol-gel [21], spray pyrolysis [22], magnetron sputtering [23], chemical vapor deposition (CVD) [24], atomic layer deposition [25], pulsed laser deposition (PLD) [20], etc., each of them having both advantages and limitations. PLD is a versatile technique used in the deposition of high-quality films based on ITO, AZO, indium-doped zinc oxide (IZO), Ga-doped ZnO (GZO), indium gallium zinc oxide (GIZO), $ZnO\text{-}Y_2O_3$ (YZO), the obtained TCO layers having adequate properties for optoelectronic device area [26–30].

Patterning techniques such as X-ray lithography, electron projection lithography, ion beam projection lithography, multiple e-beam lithography, extreme ultraviolet lithography, or nanoimprint lithography (NIL) are essential in the niche technology that manufactures high-volume and low-cost nanoscale devices [31–34]. The development and improvement of NIL technique have extended the nanoscale fabrication from standard semiconductor devices for electronics and optoelectronics to complex ones for optics, plasmonics, microfluidics, or biomimetic area [35–39]. Among NIL technologies, ultraviolet nanoimprint lithography (UV-NIL) is an efficient technique because it allows the manufacture of a wide range of pattern sizes and shapes on different rigid or flexible substrates [34, 40].

In this chapter, we present some of our contributions regarding the TCO layers deposited by PLD on flat and UV-NIL nanopatterned glass substrates. Therefore, metal oxides films (ITO and AZO) deposited by PLD were studied for emphasizing their potential applications in the field of optoelectronic devices such as OPVs and OLEDs.

2. Pulsed laser deposition (PLD)

Pulsed laser deposition (PLD) is a well-established method used to grow thin films from a wide range of materials, enabling a stoichiometric transfer of these. Although PLD was introduced in 1965, it was applied intensively in the late 1990s [41, 42]. PLD is a physical vapor deposition technique where an external high-power laser (typically an UV laser source) ablates a target based on a single or a combination of compounds depending on the desired composition of the film [43]. In comparison with other deposition methods such as sputtering, molecular beam epitaxy, chemical vapor deposition, or thermal evaporation, PLD has the following advantages: (i) any type of substrate can be used for depositing thin films; (ii) by using UV laser sources, a wide range of materials can be ablated; (iii) the pressure during the deposition process can be choose from 10^{-7} mbar up to 1 mbar; (iv) due to progressive growth with each laser pulse, a rigorous control of the thickness is possible; (v) the stoichiometry can be preserved or changed in a controlled manner during the deposition; (vi) the kinetic energy of the evaporated species can be moderated in order to control the film growth properties; (vii) a background gas can be used in order to obtain the adequate reactive atmosphere; (viii) multilayered thin films can be obtained by switching different target materials in the deposition cycle; and (ix) assure the purity of the initial composition because the ablation source is the light [42–45]. As any deposition technique, the PLD process has also some drawbacks: (i) limited deposition area for standard setups; (ii) the uniformity of the deposition is influenced by energy profile and inhomogeneity of the laser pulse; (iii) macroscopic and microscopic droplets are sometimes ejected from the target [45, 46].

PLD is a versatile method that proved its potential in different research areas considering that a wide class of the materials can be ablated using excimer lasers and deposited as thin films [42, 44, 47–53]. Thus, metal films, semiconductor films, superconductors, ceramic layers, oxides, insulators can be easily obtained by this laser technique [54, 55]. Moreover, nanostructures with different morphologies such as nanowires, nanoflowers, nanorods, nanotubes, and even quantum dots based on ZnO, ITO, graphene, molybdenum disulfide (MoS_2), tungsten disulfide (WS_2), cadmium selenide (CdSe) can be deposited by PLD [45, 47, 56–58]. The thin films or nanostructures fabricated by PLD were integrated in various devices: photovoltaics, environmental sensors, actuators, light emitters, ferroelectrics, photocatalysis, biomaterials, medical implants, etc. [45, 47, 59].

A common PLD deposition setup is depicted in **Figure 1**. Hence, the growth of the thin film is the result of the interaction between the laser beam and the target. When the laser fluence (the energy delivered per unit area at given pulse duration) reaches the ablation threshold, the vaporization of the material from the target surface takes place, process followed by the generation and expansion of the plasma plume. Further, the plasma species (free electrons, ions, neutral atoms, molecules) with appropriate energy nucleates on the deposition support [45, 59, 60]. In PLD, the film growth and the film quality depend generally on various experimental parameters:

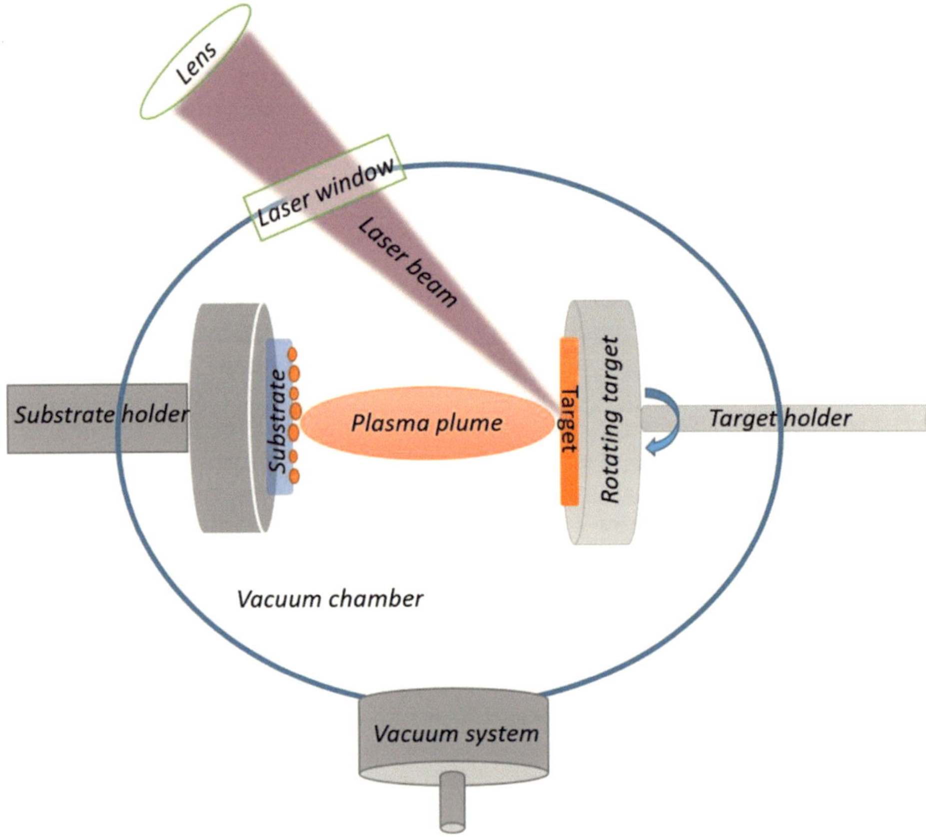

Figure 1.
Schematic representation of PLD deposition chamber.

laser fluence, laser wavelength, pulse duration, repetition rate, target-substrate distance, background gas and its pressure, quality of the target substrate temperature, etc. Because the influence of each deposition parameter on the properties of films deposited by PLD, from specific materials, was extensively discussed and analyzed in literature, in the following we briefly resumed their importance [42, 44, 47–53].

The laser fluence is one of the principal parameters because it impacts the kinetic energy of the species presented in the plasma plume and their movement toward the deposition substrate [52]. As was discussed by Schou, the chosen laser fluence must be high enough to induce target ablation but not so high to avoid the re-sputtering and possible implantation of some species in the film [53].

The laser wavelength is connected with the energy absorbed by the target material [61], thinner films being obtained when the target material is transparent to the laser wavelength used during the deposition. Lower threshold fluences and also low ablation rates are obtained when short laser wavelengths are used [48]. Thus, the laser wavelength must be selected depending on the material type intended to be deposited.

The pulse duration parameter can be controlled to prepare films with expected performances. In general, nanosecond pulse lasers are implied in the PLD deposition [48]. When long laser pulses are implied, the absorbed laser energy firstly heats the target surface to the melting point, and afterward at the vaporization temperature, the thermal wave penetrates the target and produces the melting of the material, evaporation appearing from the liquid phase. In the case of the femtosecond-pulse lasers, the vapor and plasma phases appear quickly, therefore the heat conduction is negligible, and as a consequence, the liquid phase is absent [62].

The pulse repetition rate influences the deposition rate, this being related to the duration necessary to get a specific thickness of the film [63]. The number of the particles, which are found as islands, grown firstly on the deposition substrate, subsequently tend to diffuse and aggregate depending on the pulse repetition rate, a higher density of islands being favored by the increase of this parameter. Moreover, it was emphasized that using higher pulse frequencies, a high density of small-size islands can be obtained facilitating the diffusion of some adatoms from islands top to the substrate, in this way films characterized by a smooth surface being obtained. At lower pulse frequencies, a low density of islands is formed resulting in rougher surfaces [64].

Although some PLD films can be fabricated just in ultrahigh vacuum, most of them required a background gas; this parameter affects the plume dynamics and furthers the growth and properties of the films [52, 65]. The background gas decreases the kinetic energy of the species presented in the plasma plume, a high pressure of this can decrease the sputtering of the film, but at the same time can lead to the preferential diffusion of some species to the deposition support [53, 66]. Argon, helium, or nitrogen is frequently used in the PLD deposition, but the most studied gas is still oxygen, due to the possibility of producing films with controlled oxygen content [50].

The target-substrate distance influences the mass ratio of the species that reach the substrate, thus influencing the thickness of the obtained film. A higher distance is equivalent with a reduction of the deposited material while a lower distance has as effect a rebound of the species due to their high kinetic energies [67]. Thus, it is essential to choose an optimal target-substrate distance. Some studies show that TCO layers on flexible substrate characterized by cracks or peeling off are obtained when the deposition is performed at lower target-substrate distance (4 cm) while cracks-free, smoother films are obtained at higher target-substrate distances (6 or 8 cm) [26].

The substrate temperature can influence the film growth and its surface morphology [67]. Even if the deposition can be carried on at room temperature leading usually to amorphous films, it was highlighted that at higher substrate temperatures, the adatom mobility increased resulting in crystalline films [52, 67]. When the temperature of the deposition substrate is increased, even the low kinetic energy species can be capable of constituting uniform layers [47].

Accordingly, the optimal PLD deposition conditions for developing high-quality complex films from a large number of materials can be found by tuning the experimental parameters involved in this laser process [50, 67].

3. Ultraviolet nanoimprint lithography (UV-NIL)

Nowadays, the transition from millimeter to micro and further to nano dimensions, the tendency to pass from rigid to flexible electronics, and also the continuous need of device enhanced efficiencies based on surface patterning using the principles of the plasmonic and photonic theories have forced the industry to search nanopatterning techniques that can be used in volume manufacturing [68]. In order to gain the industrial attention, these patterning techniques need to fulfill at least some key attributes such as: (i) high resolution; (ii) ability to simultaneously pattern different types of structures; (iii) high throughput and low defectivity; and (iv) reduced costs [69].

Under the name "NIL" can be found the classical thee imprint techniques: micro-contact printing (μ-CP), hot-embossing (also known as thermal NIL), and UV-NIL, but also the newly added roll imprint process, laser-assisted direct imprint, reverse imprint lithography, substrate conformal imprint lithography, ultrasonic NIL [32]. As a general definition, the nanoimprint lithography can be understood as a physical pressing process to replicate the master patterns into a polymer negative resist by thermal or ultraviolet curing [38]. Master is the name of the so called "mother"

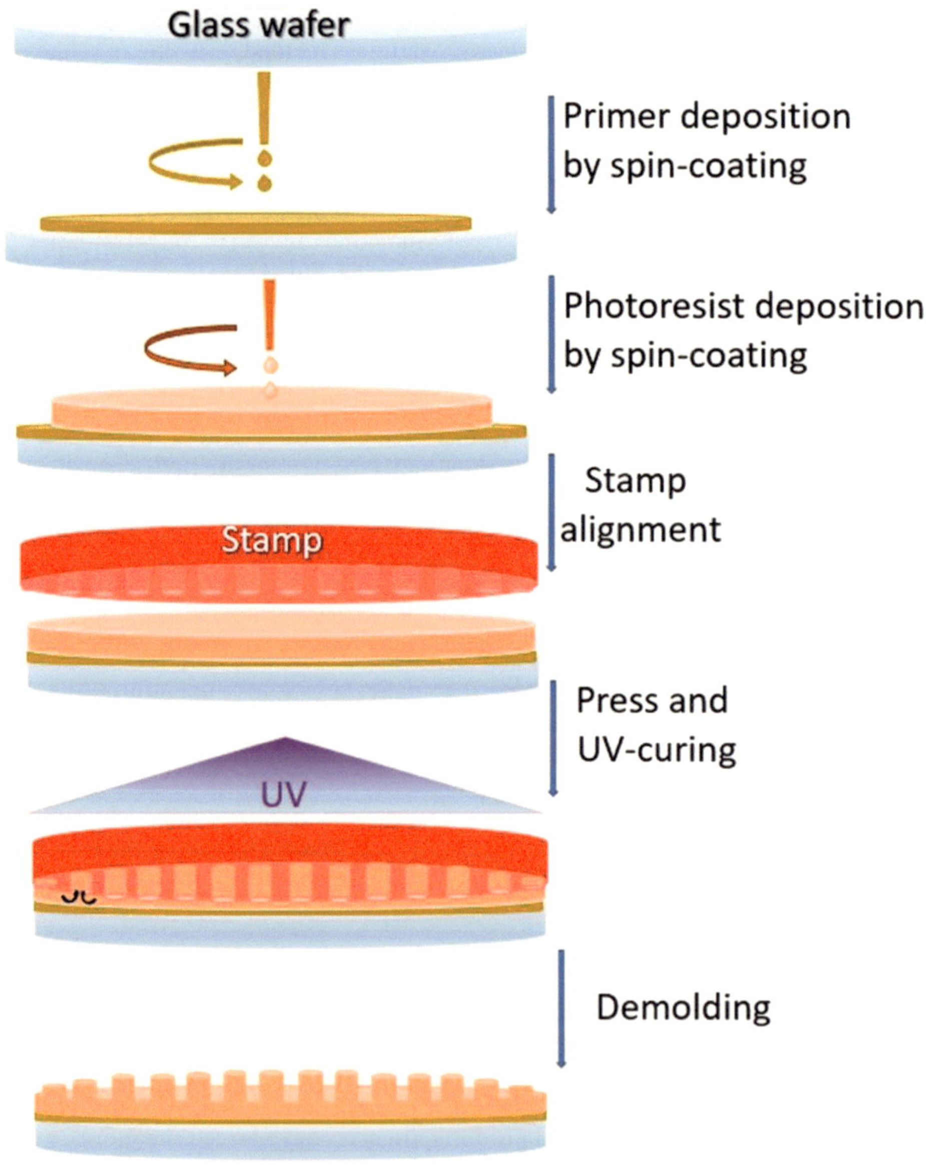

Figure 2.
Schematic representation of UV-NIL process.

template that is usually fabricated using electron beam lithography on silicon substrates. From this master, in the case of UV-NIL, rigid or soft stamps (negative copies of the master pattern designs) based on elastomeric materials can be manufactured. Thus, common materials based on silicone polymers (usually modified formulas of polydimethylsiloxane), polyimides, or polyurethanes are applied as free-standing membranes or attached to a flexible or rigid backplane [33, 37, 38, 70]. Actually, these cheaper manufactured stamps are used in the lithography process reducing the production costs and thus prolonging the lifetime of the master, this being fabricated by more time-consuming and expensive methods.

The steps involved usually in the UV-NIL process are presented in **Figure 2**. Relatively simple, they can be described as follows: (i) spin-coating deposition of both primer and photoresist on the desired substrate, each followed by a heat treatment; (ii) alignment of the stamp with the coated substrate; (iii) adding them in contact, pressing and irradiating them with UV radiation; and (iv) detaching the mask after UV curing.

The advantages of using NIL in comparison to other photolithography techniques are arising from the fact that using a direct contact between the stamp and the coated substrate, the resolution is given by the resolution of the patterns existing on the surface of stamp, which can be beyond the diffraction limits or beam scattering. However, exactly this advantage can easily become the disadvantage of the technique due to the resist filling rheology behavior and demolding capabilities [32, 33]. Therefore, one of the common defect mechanisms that appear in the NIL processes is connected with the detachment of the stamp after resist curing, when the polymer may stick on the stamp surface due to the interfacial forces (adhesion and friction forces) that appear between the resist and the stamp material. Interfacial forces are strongly linked to the quality of the stamp (design, roughness, antisticking layer, and material type), to the resist material and to the residual stress that appears during the UV irradiation due to the shrinkage of the resist that makes the stamp to adhere more to the resist surface. Taking into account all these aspects, a special attention must be paid to the selection of the materials and the process parameters that must be optimized in function of the stamp characteristics and pattern design [71, 72].

4. Indium tin oxide (ITO) and aluminum-doped zinc oxide (AZO) films deposited by PLD on flat and nanopatterned glass substrates

ITO is the most widely used TCO due to its exceptional properties, a large number of papers being focused on it [73–76]. Several works reported on the PLD deposition of ITO films and on the correlation between the experimental parameters and their optical, structural, morphological, and electrical properties, some results being well summarized by Yap and Kim [47, 77, 78]. The best properties achieved for the ITO films deposited by PLD had over 90% transparency and 7.2×10^{-5} Ωcm electrical resistivity [18].

In the last decade, many attempts were made to replace ITO due to the indium sources depletion [79]. An adequate alternative for ITO seems to be AZO, a nontoxic material that can be found at low cost—its precursors being abundant compounds, and already successfully applied in the OPV and OLED areas [80]. AZO transparent films characterized by an adequate electrical resistivity were deposited by different methods on both rigid and flexible substrates, proving its compatibility for wearable electronics [20, 81–83]. PLD technique was also used in the deposition of AZO

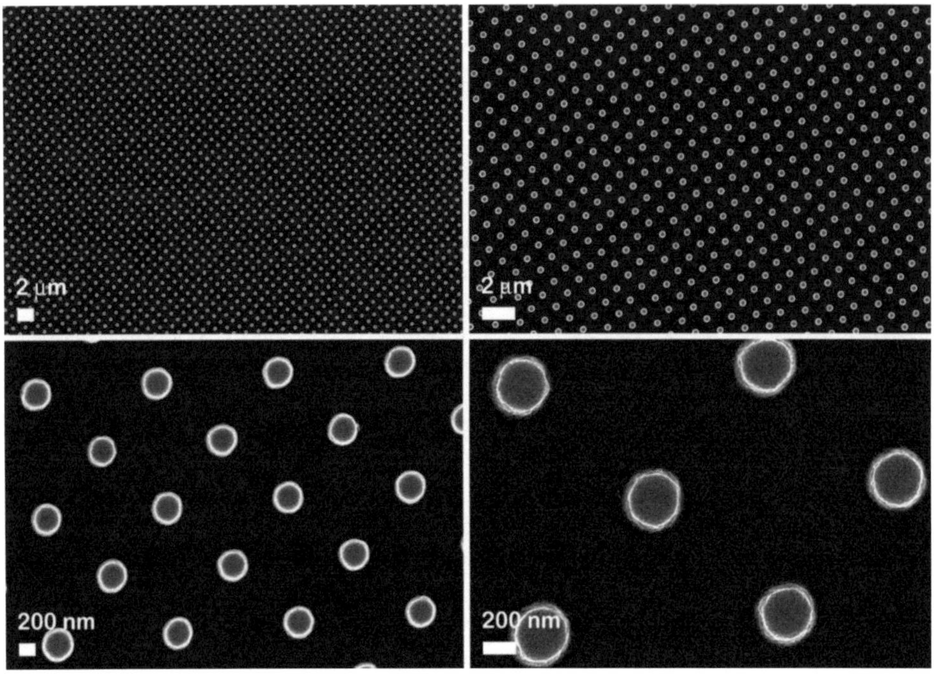

Figure 3.
FESEM images (at different magnifications) of the periodic pillars array obtained by UV-NIL method on glass substrates.

layers on either rigid glass or plastic substrates with suitable optical and electrical properties [20, 26, 84].

In the following part, the preparation steps implied in the fabrication of ITO and AZO films by PLD on flat and UV-NIL nanopatterned substrates will be described [85, 86]. The patterns were fabricated on glass by UV-NIL (EVG 620 mask aligner) using the following procedure: (i) preheating of the glass substrate for 2 min at 150°C; (ii) spin coating of a primer to enhance the adherence of the polymeric photoresist film; (iii) deposition by spin coating of the UV-resist film that further is thermally treated for 30 s at 120°C; (iv) pressing the soft stamp (mold) with the pattern model over the photoresist film with an uniform contact pressure (100 mbar); (v) exposure of the photoresist layer at UV light for 90 s; and (vi) removal of the soft mold [87]. As can be seen in the field emission scanning electron microscopy (FESEM) images from **Figure 3**, a periodic array of pillars having ~350 nm in diameter and ~ 1100 nm distance between pillars were fabricated on glass substrate by this procedure. The height of the pillars was estimated at ~250 nm from the cross-sectional FESEM images given in **Figure 4**. The quality of the patterns (height, diameter, distance between pillars) imprinted onto photoresist depends on the experimental conditions mentioned above in the UV-NIL process.

Further, TCO layers were deposited on both flat and UV-NIL patterned glass substrates by a PLD system using an excimer laser with KrF (248 nm wavelength, 25 nm pulse duration, COMPex-Pro 205, Coherent Inc.) [85, 86]. The TCO solid targets (SCI Engineered Materials) were formed by In_2O_3:SnO_2 = 90%:10% weight (ITO) and ZnO doped with 2% Al (AZO), the laser beam being directed on the target surface with a MgF_2 lens having 300 mm focal length placed outside of the deposition chamber. During the deposition, the solid targets were rotated to avoid their local

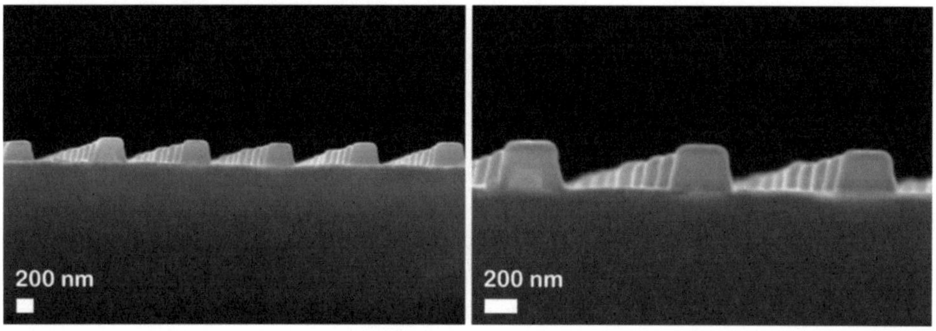

Figure 4.
Cross-sectional FESEM images (at two magnifications) of nanopatterned glass substrates.

damage. For comparison, both types of substrates were coated with TCO layers in the same deposition cycle.

The ITO solid target placed at 5 cm distance toward substrate holder was irradiated with 7000 pulses under 45° incidence angle, the laser working at 10 Hz repetition rate into a deposition chamber filled with oxygen 6.0 at 1.5 Pa pressure and working with a low laser fluence of 1.2 J/cm^2 [85]. The oxygen pressure was selected in order to obtain a low electrical resistivity, at room temperature (RT), as was mentioned in the reference [78]. The ITO layer thickness was estimated at ~340 nm as average media between the measurements made (with a profilometer) in three different points on the film deposited on flat glass substrate.

The AZO solid target placed at 8 cm distance toward substrate holder was ablated with 8000 laser pulse, a laser fluence of 2 J/cm^2, and an oxygen pressure of 1 Pa [86], the values being selected based on other preliminary results where films characterized by a high transmittance were fabricated using these experimental conditions [84]. The AZO layer thickness was estimated at ~300 nm from the interference fringes observed in the UV-VIS spectra considering two consecutive maxima and minima and the refractive index = 1.8 for AZO film with 2% Al content [88].

The TCO layers deposited by PLD were labeled taking into account the substrates type, flat (glass) or nanopatterned (NP-glass), as follows: ITO/glass, AZO/glass and ITO/NP-glass and AZO/NP-glass. The morphology and optical properties of the samples were investigated by field emission scanning electron microscopy (FESEM, Zeiss Merlin Compact field emission scanning electron microscope), atomic force microscopy (AFM, Nanonics Multiview 4000), and UV-VIS spectroscopy (Carry 5000 Spectrophotometer).

The FESEM images from **Figure 5** disclose that the ITO/glass (**Figure 5 left**) has a smooth surface while the AZO/glass (**Figure 5 right**) has a granular morphology, some particles being also presented on the surface of this sample. The results are in accordance to data already reported for ITO and AZO layers deposited by PLD [18, 84] or by other deposition techniques [89, 90].

The AFM topographic images from **Figure 6** were collected on ITO/glass (**Figure 6 left**) and AZO/glass (**Figure 6 right**), interpolated root mean square (RMS) having a low value in both cases, 1 nm and 2.8 nm, respectively. As was expected, in the case of ITO film, the RMS value is lower in comparison with ITO layers deposited by other techniques but in agreement with those calculated for ITO layers previously deposited by PLD [23, 78, 91]. It has to be mentioned that in the PLD, the resultant smooth surface is associated to the low energy density implied in the deposition process [91]. As was already

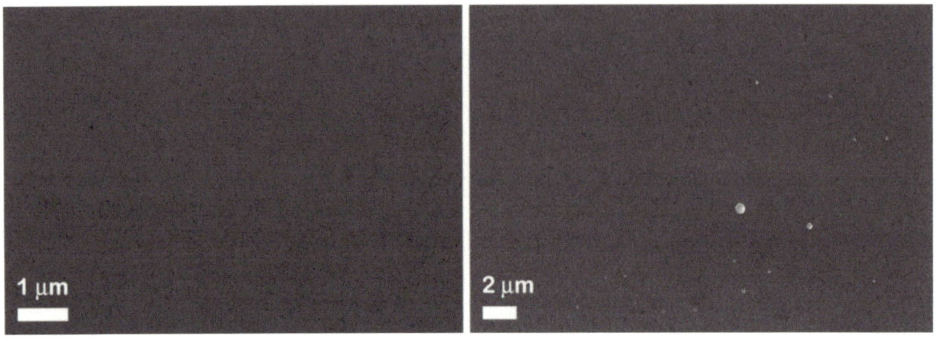

Figure 5.
FESEM images of ITO (left) and AZO (right) films deposited by PLD on flat glass substrates.

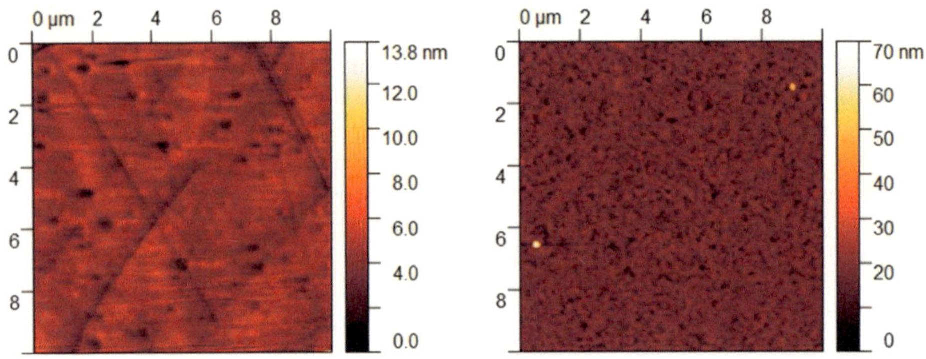

Figure 6.
AFM topographic images of ITO (left) and AZO (right) films deposited by PLD on flat glass substrates.

emphasized, such low roughness value is necessary for various applications being known that this parameter has a significant influence on optical and electrical properties of the deposited films. Thus, it was demonstrated that using a low laser fluence, ITO and AZO films characterized by a small roughness can be obtained by PLD at room temperature, making them suitable and compatible with flexible (plastic) substrates.

Analyzing the FESEM images of the ITO/NP-glass and AZO/NP-glass from **Figures 7** and **8**, respectively, it can be clearly seen that the patterns imprinted onto glass substrate are preserved during the TCO deposition by PLD. Considering that the TCO films are relatively thin (ITO ~ 340 nm and AZO ~300 nm), they tend to copy the topography of the substrate.

However, attention must be paid when the TCO layers are deposited on a patterned surface by PLD because the interaction between the ablated species, presented in the plasma plume, characterized by high kinetic energy and the deposition substrate can affect the growth of the film during the laser deposition [35, 92]. Thus, point defects can be formed due to species kinetic energy transfer toward the surface atoms [92]. In the PLD deposition on nanopatterned substrates, the first encountered layer is that based on photoresist (polymer) nanopillars. Nevertheless, the pillars are clearly observed in the FESEM images of the TCO deposited of nanopatterned glass substrates, only a small change in their shape being noted (in the case of ITO/NP-glass from cylindrical into a pyramid trunk-like one). Both TCO films seem similar at lower

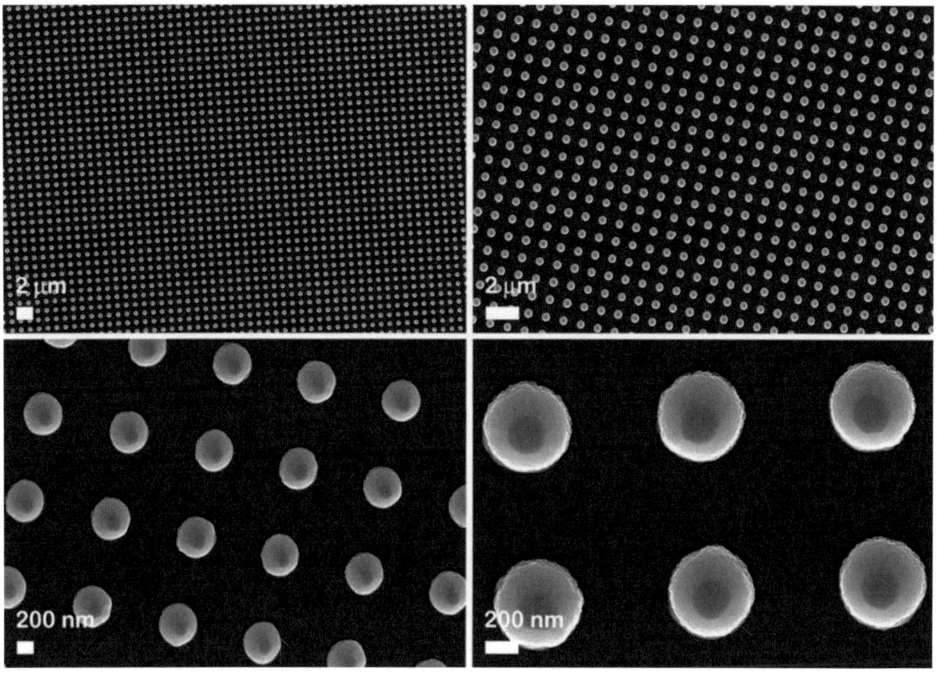

Figure 7.
FESEM images (at different magnifications) of ITO films deposited by PLD on nanopatterned glass substrates.

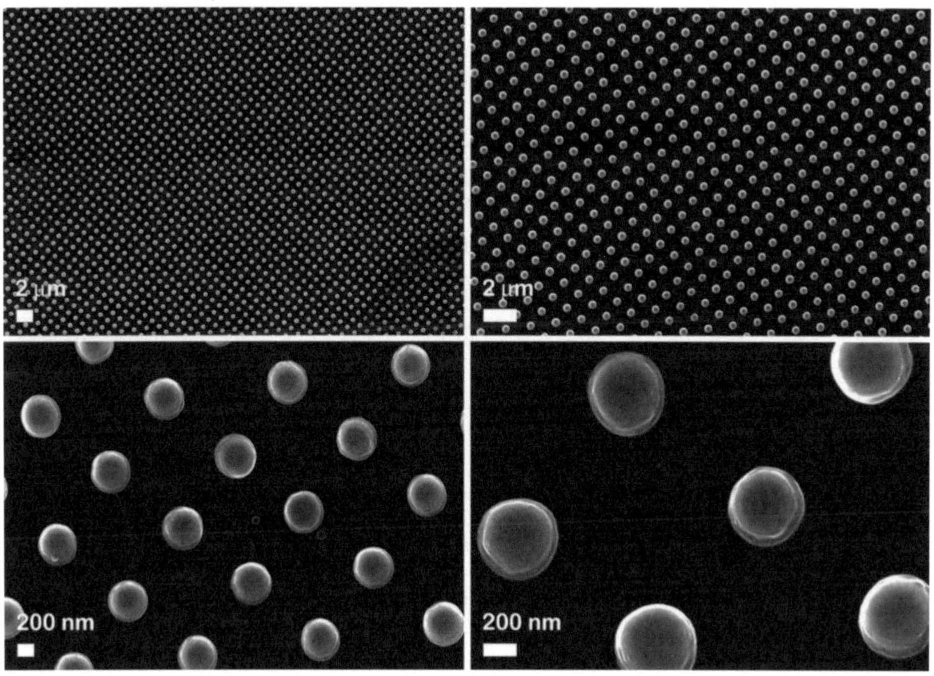

Figure 8.
FESEM images (at different magnifications) of AZO films deposited by PLD on nanopatterned glass substrates.

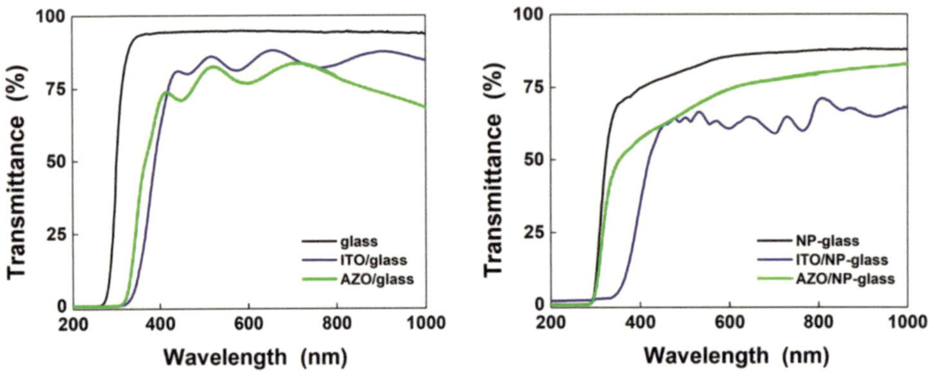

Figure 9.
UV-VIS spectra of TCO layers (ITO or AZO) deposited by PLD on flat (left) and nanopatterned (right) glass substrates.

magnification, some differences due to the film thickness and the specific morphology being visible only at higher magnification. Thus, in comparison to the nanopatterned glass substrates, an enlargement in the pillars width and a narrowing in the distance between pillars are remarked, the TCO films tending to fill the space between pillars. Although the TCO films have thickness appropriate to the pillars' height, these are not hidden by the deposited layers.

The optical transmittance is an essential criterion for the selection of the TCO films for their use in the field of OPV and OLED. Hence, the UV-VIS spectra of the prepared samples were presented in **Figure 9**. The TCO layers deposited on flat glass substrates are characterized by a transmittance over 80% for ITO and 75% for AZO in the visible part of the solar spectrum. Interference maxima are visible for both analyzed materials, their presence being associated with the uniformity of the deposited films [23]. This is not surprising, as it is known that high-quality layers can be obtained by PLD [93]. The refractive index (n) value was estimated from the interference maxima and minima with the Swaneopel method [94] as being ~1.9 for ITO/glass, value characteristic for this TCO films deposited by PLD. Kim reported that the n value depends on the Sn concentration in the targets and on the substrate temperature, for example, ITO films with low n were obtained increasing the substrate temperature [78]. Additionally, the band gap value was estimated from the UV-VIS spectra. Thus, for ITO/glass, the band gap was estimated at ~3.65 eV, in agreement with other results reported for ITO layers deposited by PLD or by magnetron sputtering, other technique frequently used to deposit commercially available ITO layers, with the same SnO_2 content (10%) [23, 91]. It is well known that this oxide is a semiconductor characterized by a direct wide band gap that can exceed 3.0 eV [95]. According to the Burstein-Moss theory, the increase in the band gap is linked to the increase in the film's carrier concentration [78].

Compared with the ZnO band gap value (3.3 eV [96]), the AZO/glass band gap was estimated at ~3.7 eV, similar to the value reported for AZO grown by PLD at room temperature and 1 Pa oxygen pressure [97]. Depending on the experimental conditions, especially by the oxygen pressure and the substrate temperature, the band gap of the AZO films deposited by PLD can take value between 3.32 and 3.77 eV [98].

In the case of the TCO layers deposited on nanopatterned glass substrate, a lowering in the transmittance is noticed in the UV-VIS spectra compared with the ones deposited on flat glass substrates. Moreover, the pillars introduced additional absorptions and reflections at interfaces [35]. The light couples to waveguide modes via diffraction and thus is trapped in the nanostructures, the pattern characteristics (mainly the period) affecting the optical properties of the films deposited on it [99]. Also, a shift of the absorption edge is visible for both transparent electrodes. A possible explanation for the peculiar behavior observed in the absorption edge shift of nanopatterned TCO (ITO/NP-glass to long wavelength region and AZO/NP-glass to short wavelength region) can be linked to the arrangement of the molecules inside the cavities determined by the nanostructuration. Thus, the interaction between the neighboring molecules can modify differently the energy levels of nanopatterned TCO with effect on their band gap.

Electrical properties of the prepared TCO layers are considered key features since, in the field of optoelectronic applications, conductive films are required. Hall measurements were performed on ITO/glass and ITO/NP-glass samples in order to analyze their electrical parameters, the obtained values being presented in **Table 1**.

In principle, the electrical resistivity values of ITO films deposited on flat and nanopatterned glass substrates are lower than ~4×10^{-4} Ωcm reported for ITO films deposited at room temperature by PLD [100] in the same conditions (laser wavelength, target composition, and repetition rate) with those used in our study. Interesting, the electrical resistivity value of ITO film deposited on flat glass substrate is nearly to that of ITO films deposited by PLD from targets with different SnO_2 content (5 or 10%) but with a heated substrate [18, 91, 101]. Kim carried on a comprehensive study regarding the influence of various experimental parameters such as oxygen pressure, SnO_2 content, and deposition temperature on the resistivity of ITO films deposited by PLD [78]. Hence, this work shows that the resistivity of ITO film is influenced by the oxygen pressure through the number of the oxygen vacancies presented in the TCO layer. Also, the resistivity of ITO films is sensitive to the SnO_2 content, an increase up to 5% leads to the resistivity decrease while an increase above this percent results in the increase of resistivity because the concentration of the electron traps expands due to Sn excess [91].

The carrier concentration values of ITO films deposited on flat and nanopatterned glass substrates are in concordance with those reported usually on ITO films deposited by PLD [78]. The refractive index of ITO films is influenced by the carrier density, a reduction of this parameter being possible by increasing the electron density, which can be achieved by enlarging the Sn content from the deposition target up to a certain value [78].

The extracted Hall mobility values of ITO films deposited on flat and nanopatterned glass substrates are just a little smaller than other value reported for ITO films deposited by PLD [91] utilizing the same deposition target with that implied in our work. The low Hall mobility values of ITO films can be related to the carrier-carrier scattering [44].

Sample	ITO/glass	ITO/NP-glass
Resistivity (Ωcm)	1.8×10^{-4}	2.8×10^{-4}
Mobility (cm^2/Vs)	10.6	15.1
Carrier concentration (cm^{-3})	3.3×10^{21}	1.5×10^{21}
Sheet resistance (Ω/sq)	5.3	8

Table 1.
Electrical parameters of ITO films deposited by PLD on flat and nanopatterned glass substrates evaluated from Hall investigations.

In the case of AZO film deposited on flat glass substrates, the resistivity was evaluated to be 2.4 × E^{-4} Ωcm using a Jandel four-point probe, the value being in the same range with others obtained for the AZO layers deposited by PLD on glass substrates [20, 102] using the same oxygen pressure with that applied in our study. A thoroughgoing study regarding the influence of the oxygen pressure on the optical and electrical properties of some AZO layers deposited by PLD was carried on in Ref. [102] pointing out that the films grown at a low oxygen pressure (under 3 Pa) have a compact structure characterized by a low resistivity.

5. TCO layers deposited by PLD on flat and nanopatterned glass substrates for developing organic heterostructures

The TCO films (ITO and AZO) deposited by PLD on flat and nanopatterned glass substrates were used for developing organic heterostructures for optoelectronic applications. Schematic representation of two organic heterostructures and their I-V characteristics are given in **Figure 10**: one based on adenine (Ade), the nucleic acid base film being deposited on ITO by vacuum thermal evaporation [103], and another based on N,N′-di(1-naftalenil)-N,N′-diafenil-(1,1′-bifenil)-4,4′-diamina (α-NPD), 1,4-bis [4-(N,N-diphenylamino)phenylvinyl] benzene (P78) and 4,7 diphenyl-1,10-phenanthroline (BPhen), the three stacked organic films being deposited on AZO by matrix-assisted pulsed laser evaporation (MAPLE) [86]. For both organic structures, aluminum electrode (100 nm) was deposited by vacuum thermal evaporation.

Hence, in the case of adenine deposited on ITO/glass substrate, the I-V characteristic (recorded in dark between −1 V and 1 V applied voltage) is changed from linear (at small voltage) to nonlinear at higher voltage (>0.5 V) probably due to the different properties shown by the contacts ITO/adenine and adenine/Al [103]. Regarding

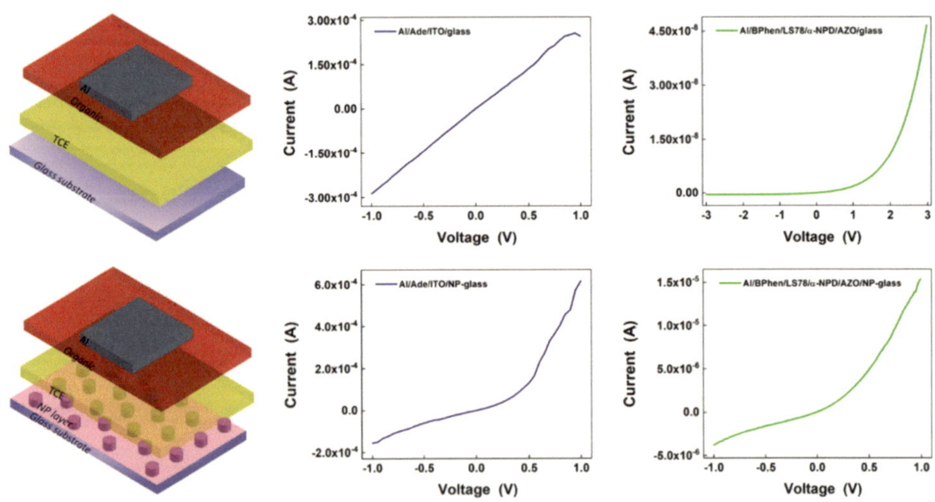

Figure 10.
Schematic representation of the organic structures using TCE deposited by PLD on flat and nanopatened glass substrates and I-V characteristics recorded on representative organic structures (single organic film – blue curve and three stacked organic films – green curve).

the electrode patterning, it is expected that this effect induces some changes in the electrical properties of the investigated structures by modifying the electrical field, which in turn can affect the charge carrier transport and their collection [104]. The scattering/recombination processes can be influenced by (i) the enlargement of the contact area between the nanopatterned TCO and the organic film, (ii) the change in the pathway of the charge carriers to the electrodes due to the presence of pillars; and (iii) the morphology of films characterized by grain boundaries. Compared with the structure prepared on ITO/glass electrode, the shape of the I-V characteristic of the structure deposited on ITO/NP-glass electrode was changed into a very close rectifying diode behavior. At small voltage, a slow increase in the current value is noted at the same time with the voltage increase, while a faster increase in the current is obtained after 0.5 V probably due to the growth of the number of electrons that cross the barrier and are more easily collected to the patterned electrode [103].

Concerning N,N′-di(1-naftalenil)-N,N′-diafenil-(1,1′-bifenil)-4,4′-diamina, 1,4-bis [4-(N,N-diphenylamino)phenylvinyl] benzene and 4,7 diphenyl-1,10-phenanthroline, an OLED-type structure was practically obtained using a hole transport layer (α-NPD), an emissive film (P78), and an electron transport layer (BPhen), respectively. Hence, the I-V characteristic plotted for the structure prepared on AZO/glass electrode presents a diode behavior. The structure fabricated on AZO/NP-glass electrode evidenced an improvement in the current value (at 1 V), meaning that the electrode patterning influences positively the electrical properties of the organic structures obtained on it [86], the charge transport being favored by the enlargement of the contact area between the nanopatterned AZO and the organic films [35]. This improvement recorded in the current value could be reflected in the final performances of the organic device fabricated on this type of nanostructured TCO.

Consequently, the optical and electrical properties of the organic structures fabricated on nanopatterned transparent electrodes can be enhanced due to the nanopatternation process. Taking into consideration that the organic heterostructures developed on TCO substrates are already part of our daily life (Heliatek company develops projects based on OPV solar films that can be attached in different locations or building facades or roofs [105], and LG Display produces OLED TV panels offering its OLED panels to other companies such as LG Electronics, Sony, Vizio, and Panasonic [106]), the organic layers deposited on patterned TCO can be also applied in the field of the organic optoelectronic devices.

6. Conclusions

TCO films (ITO and AZO) were deposited by PLD on flat and UV-NIL nanopatterned glass substrates, further these being used for developing organic heterostructures, which can find applications in optoelectronic device area. Thus, the glass substrates were patterned by UV-NIL technique, nanopillars arrays with suitable dimensions (width ~350 nm, height ~250 nm, and separation step(pitch) ~1100 nm) being fabricated. Although, the magnetron sputtering is preferred as deposition technique on large substrates, PLD is a viable alternative for fabricating high-quality TCO films with reduced roughness and appropriate optical and electrical properties by tuning the experimental deposition parameters such as: substrate temperature, oxygen pressure, target content, and laser fluence. Moreover, because the deposition of TCO films was carried at room temperature and the obtained TCO layers are characterized by low electrical resistivity, this laser technique can be also applied in the TCO deposition on plastic substrates for developing flexible devices.

The investigations prove that AZO is suitable for replacing ITO in TCO domain considering that the deposited AZO layers are featured by similar optical and electrical properties to those revealed by ITO layers.

Organic heterostructures were deposited on the fabricated TCO films (ITO and AZO) by vacuum thermal evaporation or matrix-assisted pulsed laser evaporation. The electrical measurements show that the patterning effect improves the optical and electrical properties of the organic heterostructures obtained on the TCO layers. Consequently, compared with an organic structure developed on a flat TCO electrode, an organic structure fabricated on a nanopatterned TCO electrode can be more efficient in the optoelectronic device area.

Acknowledgements

This research was funded by the Romanian Ministry of Research, Innovation and Digitization through the National Core Program PN19-03 (contract no. 21 N/2019) and PN-III-P4-IDPCE-2020-1691 (contract no. 66/2021).

Conflict of interest

The authors declare no conflict of interest.

Author details

Marcela Socol*, Nicoleta Preda, Carmen Breazu and Oana Rasoga
National Institute of Materials Physics, Magurele, Romania

*Address all correspondence to: marcela.socol@infim.ro

References

[1] Shen JJ. Recently-explored top electrode materials for transparent organic solar cells. Synthetic Metals. 2021;**271**:116582. DOI: 10.1016/j.synthmet.2020.116582

[2] Jeong EG, Kwon JH, Kang KS, Jeong SY, Choi KC. A review of highly reliable flexible encapsulation technologies towards rollable and foldable OLEDs. Journal of Information Display. 2020;**21**:19-32. DOI: 10.1080/15980316.2019.1688694

[3] Wan J, Xia Y, Fang J, Zhang Z, Xu B, Wang J, et al. Solution-processed transparent conducting electrodes for flexible organic solar cells with 16.61% efficiency. Nano-Micro Letters. 2021;**13**:44. DOI: 10.1007/s40820-020-00566-3

[4] Jin K, Xiao Z, Ding L. 18.69% PCE from organic solar cells. Journal of Semiconductors. 2021;**42**:060502. DOI: 10.1088/1674-4926/42/6/060502

[5] Narasimhan VK, Cui Y. Nanostructures for photon management in solar cells. Nano. 2013;**2**:187-210. DOI: 10.1515/nanoph-2013-0001

[6] Tang Z, Tress W, Inganäs O. Light trapping in thin film organic solar cells. Materials Today. 2014;**17**:389-396. DOI: 10.1016/j.mattod.2014.05.008

[7] Tadeson G, Sabat RG. Enhancement of the power conversion efficiency of organic solar cells by surface patterning of azobenzene thin films. ACS Omega. 2019;**4**:21862-21872. DOI: 10.1021/acsomega.9b02844

[8] Amalathas PA, Alkaisi MM. Nanostructures for light trapping in thin film solar cells. Micromachines. 2019;**10**:619. DOI: 10.3390/mi10090619

[9] Yu JW-C, Guo Y-B, Chen J-Y, Hong FC-N. Nano-imprint fabrication and light extraction simulation of photonic crystals on OLED. Proceeding of SPIE. 2008;**7140**:71400C. DOI: 10.1117/12.806888

[10] Huang YH, Lin K-C, Zeng X, Sarma M, Ni F, Shiu Y-J, et al. High-efficiency organic light emitting diodes using high-index transparent electrode. Organic Electronics. 2020;**87**:105984. DOI: 10.1016/j.orgel.2020.105984

[11] Shi XB, Qian M, Wang ZK, Liao LS. Nano-honeycomb structured transparent electrode for enhanced light extraction from organic light-emitting diodes. Applied Physics Letters. 2015;**106**:223301. DOI: 10.1063/1.4922040

[12] Madigan CF, Lu MH, Sturm JC. Improvement of output coupling efficiency of organic light-emitting diodes by backside substrate modification. Applied Physics Letters. 2000;**76**:1650-1652. DOI: 10.1063/1.126124

[13] Park CH, Kang SW, Jung SG, Lee DJ, Park YW, Ju B-K. Enhanced light extraction efficiency and viewing angle characteristics of microcavity OLEDs by using a diffusion layer. Scientific Reports. 2021;**11**:3430. DOI: 10.1038/s41598-021-82753-9

[14] Anand A, Islam MM, Meitzner R, Schubert US, Hoppe H. Introduction of a novel figure of merit for the assessment of transparent conductive electrodes in photovoltaics: Exact and approximate form. Advanced Energy Materials.

2021;**11**:2100875. DOI: 10.1002/aenm.202100875

[15] Minami T. Transparent conductive oxides for transparent electrode applications. In: Svensson BG, Pearton SJ, Jagadish C, editors. Semiconductors and Semimetals. 1st ed. Amsterdam: Elsevier; 2013. pp. 159-200, ISSN 0080-8784. DOI: 10.1016/B978-0-12-396489-2.00005-9

[16] Lee HB, Jin W-Y, Ovhal MM, Kumar N, Kang J-W. Flexible transparent conducting electrodes based on metal meshes for organic optoelectronic device applications: A review. Journal of Materials Chemistry C. 2019;7:1087-1110. DOI: 10.1039/C8TC04423F

[17] Pang S, Hernandez Y, Feng X, Müllen K. Graphene as transparent electrode material for organic electronics. Advanced Materials. 2011;**23**:2779-2795. DOI: 10.1002/adma.201100304

[18] Suzuki A, Matsushita T, Aoki T, Yoneyama Y, Okuda M. Pulsed laser deposition of transparent conducting indium tin oxide films in magnetic field perpendicular to plume. Japanese Journal of Applied Physics. 2001;**40**:L401. DOI: 10.1143/JJAP.40.L401

[19] Kan Z, Wang OZ, Firdaus Y, Babics M, Alshareef HN, Beaujuge PM. Atomic-layer-deposited AZO outperforms ITO in high-efficiency polymer solar cells. Journal of Materials Chemistry A. 2018;**6**:10176-10183. DOI: 10.1039/C8TA02841A

[20] Dosmailov M, Leonat LN, Patek J, Roth D, Bauer P, Scharber MC, et al. Transparent conductive ZnO layers on polymer substrates: Thin film deposition and application in organic solar cells. Thin Solid Films. 2015;**591**:97-104. DOI: 10.1016/j.tsf.2015.08.015

[21] Ghomrani F-Z, Iftimie S, Gabouze N, Serier A, Socol M, Stanculescu A, et al. Influence of Al doping agents nature on the physical properties of Al:ZnO films deposited by spin-coating technique. Optoelectronics and Advanced Materials-Rapid Communications. 2011;**5**:247-251

[22] Rana R, Chakraborty J, Tripathi SK, Nasim N. Study of conducting ITO thin film deposition on flexible polyimide substrate using spray pyrolysis. Journal of Nanostructure in Chemistry. 2016;**6**:65-74. DOI: 10.1007/s40097-015-0177-7

[23] Prepelita P, Filipescu M, Stavarache I, Garoi F, Craciun D. Transparent thin films of indium tin oxide: Morphology–optical investigations, inter dependence analyzes. Applied Surface Science. 2017;**424**:368-373. DOI: 10.1016/j.apsusc.2017.02.106

[24] Ponja SD, Sathasivam S, Parkin IP, Carmalt J. Highly conductive and transparent gallium doped zinc oxide thin films via chemical vapor deposition. Scientific Reports. 2020;**10**:638. DOI: 10.1038/s41598-020-57532-7

[25] Zhao K, Xie J, Zhao Y, Han D, Wang Y, Liu B, et al. Investigation on transparent, conductive ZnO: Al films deposited by atomic layer deposition process. Nanomaterials. 2022;**12**:172. DOI: 10.3390/nano12010172

[26] Socol G, Socol M, Stefan N, Axente E, Popescu-Pelin G, Craciun D, et al. Pulsed laser deposition of transparent conductive oxide thin films on flexible substrates. Applied Surface Science. 2012;**260**:42-46. DOI: 10.1016/j.apsusc.2012.02.148

[27] Franklin JB, Gilchrist JB, Downing JM, Roya KA, McLachlan MA. Transparent conducting oxide top contacts for organic electronics. Journal

of Materials Chemistry C. 2014;**2**:84-89. DOI: 10.1039/C3TC31296H

[28] Beckford J, Behera MK, Yarbrough K, Obasogie B, Pradhan SK, Bahoura M. Gallium doped zinc oxide thin films as transparent conducting oxide for thin-film heaters. AIP Advances. 2021;**11**:075208. DOI: 10.1063/5.0016367

[29] Mistry BV, Joshi US. Amorphous indium gallium zinc oxide thin film grown by pulse laser deposition technique. AIP Conference Proceedings. 2016;**1731**:080035. DOI: 10.1063/1.4947913

[30] Youvanidha A, Vidhya B, Nelson PI, Kannan RR, Babu SKS. Investigation on the structural, optical and electrical properties of ZnO-Y_2O_3 (YZO) thin films prepared by PLD for TCO layer applications. AIP Conference Proceedings. 2019;**2166**:020023. DOI: 10.1063/1.5131610

[31] Thanner C, Dudus A, Treiblmayr D, Berger G, Chouiki M, Martens S, et al. Nanoimprint lithography for augmented reality waveguide manufacturing. Proceedings SPIE. 2020;**11310**:1131010. DOI: 10.1117/12.2543692

[32] Lan H, Ding Y. Nanoimprint lithography. In: Wang M, editor. Lithography. 1st ed. Rijeka: Intech; 2010. pp. 457-494. DOI: 10.5772/8189

[33] Baracu AM, Avram MA, Breazu C, Bunea M, Socol M, Stanculescu A, et al. Silicon metalens fabrication from electron beam to UV-nanoimprint lithography. Nanomaterials. 2021;**11**:2329. DOI: 10.3390/nano11092329

[34] Thanner C, Eibelhuber M. UV nanoimprint lithography: Geometrical impact on filling properties of nanoscale patterns. Nanomaterials. 2021;**11**:822. DOI: 10.3390/nano11030822

[35] Stanculescu A, Breazu C, Socol M, Rasoga O, Preda N, Petre G, et al. Effect of ITO electrode patterning on the properties of organic heterostructures based on non-fullerene acceptor prepared by MAPLE. Applied Surface Science. 2020;**509**:145351. DOI: 10.1016/j.apsusc.2020.145351

[36] Nevřela J, Kuzma A. Fabrication of PhC structures by using nanoimprint lithography and their optical properties. In: Technology Transfer: Fundamental Principles and Innovative Technical Solutions: 3nd Annual Conference; November 23, 2019; Tallinn. Estonia: Scientific Route; 2019. pp. 29-31. DOI: 10.21303/2585-6847.2019.001020

[37] Dirdal CA, Jensen GU, Angelskår H, Vaagen Thrane PC, Gjessing J, Ordnung DA. Towards high-throughput large-area metalens fabrication using UV-nanoimprint lithography and Bosch deep reactive ion etching. Optical Express. 2020;**28**:15542-15561. DOI: 10.1364/OE.393328

[38] Barcelo S, Li Z. Nanoimprint lithography for nanodevice fabrication. Nano Convergence. 2016;**3**:21. DOI: 10.1186/s40580-016-0081-y

[39] Ahmed R, Ozen MO, Karaaslan MG, Prator CA, Thanh C, Kumar S, et al. Tunable fano-resonant metasurfaces on a disposable plastic-template for multimodal and multiplex biosensing. Advanced Materials. 2020;**32**:1-11. DOI: 10.1002/adma.201907160

[40] Chen J, Zhou Y, Wang D, He F, Rotello VM, Carter KR, et al. UV-nanoimprint lithography as a tool to develop flexible microfluidic devices for electrochemical detection. Lab on a Chip. 2015;**15**:3086. DOI: 10.1039/c5lc00515a

[41] Kiyek VM, Birkhölzer YA, Smirnov Y, Ledinsky M, Remes Z, Momand J, et al.

Single-source, solvent-free, room temperature deposition of black γ-CsSnI3 films. Advanced Materials Interfaces. 2020;7:2000162. DOI: 10.1002/admi.202000162

[42] Ogugua SN, Ntwaeaborwa OM, Swart HC. Latest development on pulsed laser deposited thin films for advanced luminescence applications. Coatings. 2020;**10**:1078. DOI: 10.3390/coatings10111078

[43] Schneider CW, Lippert T. Laser ablation and thin film deposition. In: Schaaf P, editor. Laser Processing of Materials. Springer Series in Materials Science. 1st ed. Berlin: Springer; 2010. pp. 89-112. DOI: 10.1007/978-3-642-13281-0_5

[44] Eason R, editor. Pulsed Laser Deposition of Thin Films: Applications-Led Growth of Functional Materials. 1st ed. New Jersey: Wiley; 2006. p. 682. DOI: 10.1002/0470052120

[45] Masood KB, Kumar P, Malik MA, Singh J. A comprehensive tutorial on the pulsed laser deposition technique and developments in the fabrication of low dimensional systems and nanostructures. Emergent Materials. 2021;**4**:737-754. DOI: 10.1007/s42247-020-00155-5

[46] Willmott PR, Huber JR. Pulsed laser vaporization and deposition. Reviews of Modern Physics. 2000;**72**:315-328. DOI: 10.1103/RevModPhys.72.315

[47] Yap SS, Yong TK, Nee CH, Tou TY. Pulsed laser deposition of ITO: From films to nanostructures. In: Yong TK, editor. Applications of Laser Ablation—Thin Film Deposition, Nanomaterial Synthesis and Surface Modification. 1st ed. Rijeka: IntechOpen; 2016. pp. 85-102. DOI: 10.5772/65897

[48] Chrisey DB, Hubler GK. Pulsed Laser Deposition of Thin Films. 1st ed. New York: Wiley; 1994. p. 648. ISBN 0471592188/9780471592181

[49] Kuzanyan AS, Kuzanyan AA. Pulsed laser deposition of large-area thin films and coatings. In: Yang D, editor. Applications of Laser Ablation—Thin Film Deposition, Nanomaterial Synthesis and Surface Modification. London, UK: IntechOpen; 2016. pp. 149-172. DOI: 10.5772/64978

[50] Amoruso S. Plume characterization in pulsed laser deposition of metal oxide thin films. In: Pryds N, Esposito V, editors. Metal Oxide-Based Thin Film Structures Formation, Characterization and Application of Interface-Based Phenomena Metal Oxides. Amsterdam: Elsevier; 2018. pp. 133-160. DOI: 10.1016/B978-0-12-811166-6.00006-6

[51] Ashfold MNR, Claeyssens F, Fugea GM, Henleya SJ. Pulsed laser ablation and deposition of thin films. Chemical Society Reviews. 2004;**33**:23-31. DOI: 10.1039/B207644F

[52] Indrizzi L, Ohannessian N, Pergolesi D, Lippert T, Gilardi E. Pulsed laser deposition as a tool for the development of all solid-state microbatteries. Helvetica Chimica Acta. 2021;**104**:e20002. DOI: 10.1002/hlca.202000203

[53] Schou J. Physical aspects of the pulsed laser deposition technique: The stoichiometric transfer of material from target to film. Applied Surface Science. 2009;**255**:5191-5198. DOI: 10.1016/j.apsusc.2008.10.101

[54] Antoni F, Stock F. Laser engineering of carbon materials for optoelectronic applications. In: Fuccio C, La Magna A. editors. Laser Annealing Processes in Semiconductor Technology Theory, Modeling and Applications in Nanoelectronics. Woodhead Publishing

Series in Electronic and Optical Materials. Amsterdam: Elsevier; 2021. p. 293-321. DOI: 10.1016/B978-0-12-820255-5.00005-2

[55] Mao SS, Zhang X. High-throughput multi-plume pulsed-laser deposition for materials exploration and optimization. Engineering. 2015;**1**:367-371. DOI: 10.15302/J-ENG-2015065

[56] Shkurmanov A, Sturm C, Lenzner J, Feuillet G, Tendille F, de Mierry P, et al. Selective growth of tilted ZnO nanoneedles and nanowires by PLD on patterned sapphire substrates. AIP Advances. 2016;**6**:095013. DOI: 10.1063/1.4963076

[57] Nikov RG, Dikovska AO, Nedyalkov NN, Avdeev GV, Atanasov PA. Au nanostructure fabrication by pulsed laser deposition in open air: Influence of the deposition geometry. Beilstein Journal of Nanotechnology. 2017;**8**:2438-2445. DOI: 10.3762/bjnano.8.242

[58] Khalaph KA, Abdalameer NK, Mousa AQ. Study the physical properties of CdSe nanostructures prepared by a pulsed laser deposition method. AIP Conference Proceedings. 2021;**2372**:130023. DOI: 10.1063/5.0065731

[59] Kannan PK, Chaudhari S, Dey SR, Ramadan M. Progress in development of CZTS for solar photovoltaics applications. Encyclopedia of Smart Materials. 2022;**2**:681-698. DOI: 10.1016/B978-0-12-815732-9.00130-3

[60] Orava J, Kohoutek T, Wagner T. Deposition techniques for chalcogenide thin films. In: Adam J-L, Zhang X, editors. Chalcogenide Glasses. Sawston: Woodhead Publishing; 2014. pp. 265-309. DOI: 10.1533/9780857093561.1.265

[61] Nyenge RL, Sechogela PT, Swart HC, Ntwaeaborwa OM. The influence of laser wavelength on the structure, morphology, and photoluminescence properties of pulsed laser deposited CaS: Eu2+ thin films. Journal of Modern Optics. 2015;**62**:1102-1109. DOI: 10.1080/09500340.2015.1020898

[62] Chichkov B, Momma C, Nolte S, von Alvensleben F, Tunnermann A. Femtosecond, picosecond and nanosecond laser ablation of solids. Applied Physics A: Materials Science & Processing. 1996;**63**:109-115. DOI: 10.1007/BF01567637

[63] Almuslet NA, & Alsheikh YH. Effect of pulse repetition rate on the properties of pulse laser deposited SiO2 thin films. 2018 International Conference on Computer, Control, Electrical, and Electronics Engineering (ICCCEEE); 12-14 August 2018; Khartoum. Sudan: IEEE; 2018. DOI:10.1109/iccceee.2018.8515852

[64] Guan L, Zhang DM, Li X, Li ZH. Role of pulse repetition rate in film growth of pulsed laser deposition. Nuclear Instruments and Methods in Physics Research Section B: Beam Interactions with Materials and Atoms. 2008;**266**:57-62. DOI: 10.1016/j.nimb.2007.10.011

[65] Smirnov Y, Schmengler L, Kuik R, Repecaud P-A, Mi N, Zhang D, et al. Scalable pulsed laser deposition of transparent rear electrode for perovskite solar cells. Advanced Materials Technolies. 2021;**6**:2000856. DOI: 10.1002/admt.202000856

[66] Kumara A, Singh RK, Prahlad V, Joshi HC. Comparative study of laser produced Li plasma plumes from thin film and solid target. Journal of Applied Physics. 2008;**104**:093302. DOI: 10.1063/1.3014031

[67] Ojeda A, Döbeli M, Lippert T. Influence of plume properties on

thin film composition in pulsed laser deposition. Advanced Materials Interfaces. 2018;**5**:1701062. DOI: 10.1002/admi.201701062

[68] Ban YH, Bonnecaze RT. Minimizing filling time for ultraviolet nanoimprint lithography with templates with multiple structures. Journal of Vacuum Science & Technology B. 2021;**39**:012601. DOI: 10.1116/6.0000648

[69] Sreenivasan SV. Nanoimprint lithography steppers for volume fabrication of leading-edge semiconductor integrated circuits. Microsystems & Nanoengineering. 2017;**3**:1-19. DOI: 10.1038/micronano.2017.75

[70] Thanner C, Breyer R, Litterscheidt J. Optimized UV nanoimprinting processes for fabrication of high fidelity patterns. Proceedings of the International Semiconductor Conference (CAS '2020); 7-9 October 2020; Sinaia, Romani. pp. 129-132. DOI: 10.1109/CAS50358.2020.9268031

[71] Li M, Chen Y, Luo W, Cheng X. Interfacial interactions during demolding in nanoimprint lithography. Micromachines. 2021;**12**:349. DOI: 10.3390/mi12040349

[72] Tochino T, Shiotsu T, Uemura K, Yasuda M, Kawata H, Hirai Y. Impact of resist shrinkage on the template release process in nanoimprint lithography. Journal of Vacuum Science & Technology B. 2014;**32**:06FG08. DOI: 10.1116/1.4901874

[73] Ho SM. A review on thin films on indium tin oxide coated glass substrate. Asian Journal of Chemistry. 2016;**28**:469-472. DOI: 10.14233/ajchem.2016.19579

[74] Txintxurreta J, G Berasategui E, Ortiz R, Hernández O, Mendizábal L, Barriga J. Indium tin oxide thin film deposition by magnetron sputtering at room temperature for the manufacturing of efficient transparent heaters. Coatings. 2021;**11**:92. DOI: 10.3390/coatings11010092

[75] Kanneboina V, Basumatary P, Agarwal P. Influence of deposition temperature on indium tin oxide thin films for solar cell applications. AIP Conference Proceedings. 2019;**2091**:020016. DOI: 10.1063/1.5096507

[76] Aydın EB, Sezgintürk MK. Indium tin oxide (ITO): A promising material in biosensing technology. TrAC Trends in Analytical Chemistry. 2017;**97**:309-315. DOI: 10.1016/j.trac.2017.09.021

[77] Kim H, Horwitz J, Piqué A, Gilmore CM, Chrisey DB. Electrical and optical properties of indium tin oxide thin films grown by pulsed laser deposition. Applied Physics A. 1999;**69**:S447-S450. DOI: 10.1007/s003390051435

[78] Kim H, Gilmore CM, Pique' A, Horwitz JS, Mattoussi H, Murata H, et al. Electrical, optical, and structural properties of indium-tin-oxide thin films for organic light-emitting devices. Journal of Applied Physics. 1999;**86**:6451-6461. DOI: 10.1063/1.371708

[79] Kawajiri K, Tahara K, Uemiya S. Lifecycle assessment of critical material substitution: Indium tin oxide and aluminum zinc oxide in transparent electrodes. Resources, Environment and Sustainability. 2022;**7**:100047. DOI: 10.1016/j.resenv.2022.100047

[80] Ravichandran K, Jabena Begum N, Snega S, Sakthivel B. Properties of sprayed aluminum-doped zinc oxide films—A review.

Materials and Manufacturing Processes. 2016;**31**:1411-1423. DOI: 10.1080/10426914.2014.930961

[81] Dimitrov D, Tsai C-L, Petrov S, Marinova V, Petrova D, Napoleonov B, et al. Atomic layer-deposited Al-doped ZnO thin films for display applications. Coatings. 2020;**10**:539. DOI: 10.3390/coatings10060539

[82] Schulzea K, Maennig B, Leo K. Organic solar cells on indium tin oxide and aluminum doped zinc oxide anodes. Applied Physics Letters. 2007;**91**:073521. DOI: 10.1063/1.2771050

[83] Chauhan RN, TiwariN ARS, Kumar J. Development of Al-doped ZnO thin film as a transparent cathode and anode for application in transparent organic light-emitting diodes. RSC Advances. 2016;**6**:86770-86781. DOI: 10.1039/C6RA14124B

[84] Socol M, Preda N, Stanculescu A, Breazu C, Florica C, Stanculescu F, et al. Organic heterostructures deposited by MAPLE on AZO substrate. Applied Surface Science. 2017;**2007**(417):196-203. DOI: 10.1016/j.apsusc.2017.02.260

[85] Socol M, Preda N, Rasoga O, Costas A, Stanculescu A, Breazu C, et al. Pulsed laser deposition of indium tin oxide thin films on nanopatterned glass substrates. Coatings. 2019;**9**:19. DOI: 10.3390/coatings9010019

[86] Socol M, Preda N, Breazu C, Rasoga O, Stanculescu A, Popescu-Pelin G, et al. Organic heterostructures deposited by maple on patterned AZO electrodes. Digest Journal of Nanomaterials and Biostructures. 2018;**13**(4):1045-1053

[87] Breazu C, Preda N, Socol M, Stanculescu F, Matei E, Stavarache I, et al. Investigations on the properties of a two-dimensional nanopatterned metallic film. Digest Journal of Nanomaterials and Biostructures. 2016;**11**:1213-1229

[88] Besleaga C, Ion L, Antohe S. AZO thin films synthes ized by rf-magnetron sputtering: The role of deposition power. Romanian Reports in Physics. 2014;**66**:993-1001. DOI: 10.1117/12.2061186

[89] Chen Z, Li W, Li R, Zhang Y, Xu G, Cheng H. Fabrication of highly transparent and conductive indium–tin oxide thin films with a high figure of merit via solution processing. Langmuir. 2013;**29**:13836-13842. DOI: 10.1021/la4033282

[90] Kheanwong J, Rattanasakulthong W. Morphology-dependent optical transmission of rf-sputtered ZnO:Al film on glass substrate. Digest Journal of Nanomaterials and Biostructures. 2015;**10**:759-768

[91] Kim SH, Park NM, Kim T, Sung GY. Electrical and optical characteristics of ITO films by pulsed laser deposition using a 10 wt.% SnO2-doped In2O3 ceramic target. Thin Solid Films. 2005;**475**:262-266. DOI: 10.1016/j.tsf.2004.08.032

[92] Zhang L. Guan. 4.06—Laser Ablation. Saleem Hashmi, Gilmar Ferreira Batalha, Chester J. Van Tyne, Bekir Yilbas. Comprehensive Materials Processing. Elsevier; 2014. pp. 125-169. ISBN 97800809653

[93] Chaluvadi SK, Mondal D, Bigi C, Knez D, Rajak P, Ciancio R, et al. Pulsed laser deposition of oxide and metallic thin films by means of Nd:YAG laser source operating at its 1st harmonics: Recent approaches and advances. Journal of Physics: Materials. 2021;**4**:032001. DOI: 10.1088/2515-7639/abe661

[94] Swanepoel R. Determination of surface roughness and optical constants

of inhomogeneous amorphous silicon films. Journal of Physics E: Scientific Instruments. 1984;**17**:896. DOI: 10.1088/0022-3735/17/10/023

[95] Afre RA, Sharma N, Sharon M, Sharon M. Transparent conducting oxide films for various applications: A review. Reviews on Advanced Materials Science. 2018;**53**:79-89. DOI: 10.1515/rams-2018-0006

[96] Preda N, Enculescu M, Enculescu I. Polysaccharide-assisted crystallization of ZnO micro/nanostructures. Materials Letters. 2014;**115**:256-260. DOI: 10.1016/j.matlet.2013.10.081

[97] Gondoni P, Ghidelli M, Di Fonzo F, Russo V, Bruno P, Martí-Rujas J, et al. Structural and functional properties of Al:ZnO thin films grown by Pulsed Laser Deposition at room temperature. Thin Solid Films. 2012;**520**:4707-4711. DOI: 10.1016/j.tsf.2011.10.072

[98] Shan FK, Yu YS. Band gap energy of pure and Al-doped ZnO thin films. Journal of the European Ceramic Society. 2004;**24**:1869-1872. DOI: 10.1016/S0955-2219(03)00490-4

[99] Khan I, Bauch M, Dimopoulos T, Dostalek J. Nanostructured as-deposited indium tin oxide thin films for broadband antireflection and light trapping. Nanotechnology. 2017;**28**:325201. DOI: 10.1088/1361-6528/aa79df

[100] Wu Y, Marée CHM, Haglund RF Jr, Hamilton JD, Morales Paliza MA, Huang MB, et al. Resistivity and oxygen content of indium tin oxide films deposited at room temperature by pulsed-laser ablation. Journal of Applied Physics. 1999;**86**:991. DOI: 10.1063/1.370864

[101] Fang X, Mak CL, Zhang S, Wang Z, Yuan W, Ye H. Pulsed laser deposited indium tin oxides as alternatives to noble metals in the near-infrared region. Journal of Physics: Condensed Matter. 2016;**28**:224009. DOI: 10.1088/0953-8984/28/22/224009

[102] Gondoni P, Ghidelli M, Di Fonzo F, Carminati M, Russo V, Bassi AL, et al. Structure-dependent optical and electrical transport properties of nanostructured Al-doped ZnO. Nanotechnology. 2012;**23**:365706. DOI: 10.1088/0957-4484/23/36/365706

[103] Breazu C, Socol M, Preda N, Rasoga O, Costas A, Socol G, et al. Nucleobases thin films deposited on nanostructured transparent conductive electrodes for optoelectronic applications. Scientific Reports. 2021;**11**:7551. DOI: 10.1038/s41598-021-87181-3

[104] Ray B, Khan MR, Black C, Alam MA. Nanostructured electrodes for organic solar cells: Analysis and design fundamentals. IEEE Journal of Photovoltaics. 2013;**3**:318-329. DOI: 10.1109/JPHOTOV.2012.2220529

[105] Heliatek. Solar electricity—An essential part of our future [Internet]. 2022. Available from: https://www.heliatek.com/en/products. [Accessed: June 06, 2022]

[106] OLED TV: Introduction and Market News [Internet]. 2022. Available from: https://www.oled-info.com/oled-tv. [Accessed: June 06, 2022]

Chapter 9

Development and Applications of Aluminum Nitride Thin Film Technology

Cícero L.A. Cunha, Tales C. Pimenta and Mariana Amorim Fraga

Abstract

Aluminum nitride (AlN) thin films have aroused the interest of researchers due to their unique physicochemical properties. However, further studies on these semiconductor materials are still necessary to establish the manufacturing of high-performance devices for applications in various areas, such as telecommunications, microelectronics, and biomedicine. This chapter introduces AlN thin film technology that has made a wide range of applications possible. First, the main physicochemical properties of AlN, its wurtzite crystalline structure, and the incorporation of oxygen during the thin film deposition process are presented. Furthermore, the growth of AlN films by different techniques and their applications as a buffer layer and sensing layer are summarized. Special attention was given to the sputtering deposition process and the use of sputtered AlN films in SAW sensors.

Keywords: aluminum nitride, wurtzite crystalline structure, sputtering, thin film, buffer layer, sensing layer, SAW sensors

1. Introduction

Aluminum nitride (AlN) thin films have attracted much attention due to their excellent properties suitable for the manufacture of devices, which meet various applications [1–7]. In addition to their properties highlighted in **Figure 1**, these materials have been much investigated due to their high piezoelectricity and high surface acoustic velocity ($v_\perp$ = 5600 m/s and $v_\parallel$ = 11,000 m/s) [2, 3], suitable electromechanical coupling, chemical stability [3, 8–13] and good transparency in the region of the visible, infrared and ultraviolet [10, 13].

In recent years, AlN has been shown as an outstanding candidate for the development of micro-electro-mechanical systems (MEMS), particularly surface acoustic wave (SAW) devices operating in high frequency and in thermally and chemically harsh environments [3].

It has been observed that the thin film deposition technique influences the preferred orientation of the AlN film. Different techniques, such as chemical vapor deposition (CVD), molecular beam epitaxy (MBE), and reactive sputtering (RF or DC, with or

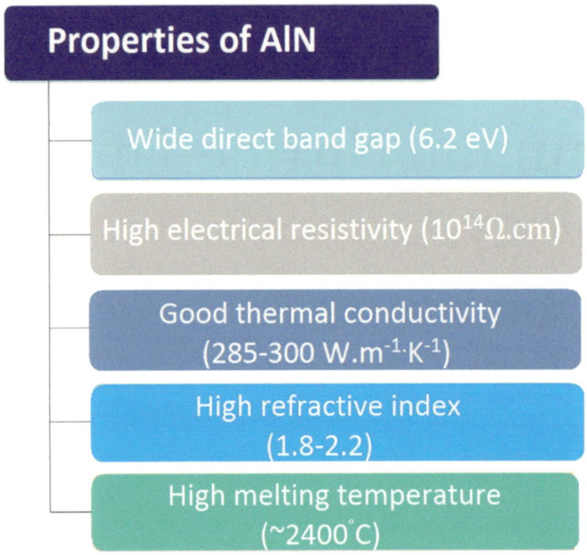

Figure 1.
Some properties of AlN.

without magnetron), have been used for the growth of AlN films [3, 12]. The advantages of RF magnetron sputtering technique are that besides the deposition parameters can be easily controlled, it uses low temperature (<400°C) and has compatibility with CMOS technology [3, 9]. Generally, the deposition vacuum systems used in industries have limitations due to the use of large vacuum chambers and of the laborious tasks of environment cleaning processes (chamber, substrates, and everything other system parts). In deposition systems used in microelectronics applications, there is also the possibility of incorporation of oxygen in the films due to the native oxygen on the silicon wafers and the residual relative to the background pressure 1.33×10^{-4} Pa ($\approx 10^{-6}$ Torr). Thus, in these conditions, it is difficult to deposit crystalline and highly oriented textural AlN (100) films with low oxygen concentration in their structure, good stoichiometry Al/N (≈1:1), and thickness close to 500 nm [14, 15].

In this chapter, an overview of the AlN thin film technology is presented. First, the structure and orientation of AlN material are described. Next, the incorporation of oxygen in this material is discussed. The largest section of the chapter is devoted to the growth of AlN films and their use as a buffer layer and sensing layer. Finally, some examples of SAW sensors based on AlN films are presented.

2. Structure and crystal orientation of the AlN material

In the non-excited state, the aluminum has three electrons in its valence layer, distributed as $3s^2 3p^1$, so it presents a sublevel s complete with two electrons and one p sublevel semi-complete with an electron and two p sublevels empty, as shown in **Figure 2a**. On the other hand, the nitrogen has five electrons in the valence layer distributed as $2s^2 2p^3$, where the s sublevel is filled with two electrons and the three sublevels p_x, p_y, and p_z are semi-complete with one electron each, as shown in **Figure 1b**. Now, in the excited state, the aluminum sublevels $3s^2 3p^1$ rise to four

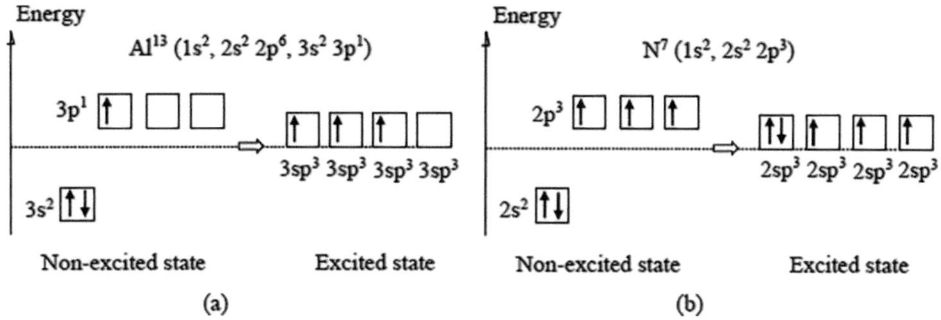

Figure 2.
Hybridization of aluminum and nitrogen. (a) Valence layer in the non-excited state and hybrid sublevels of aluminum, $3sp^3$, in the excited state. (b) Valence layer in the non-excited state and hybrid sublevels of nitrogen, $2sp^3$, in the excited state.

hybrids $3sp^3$ sublevels, in which three sublevels are semi-complete with one electron each, and the remaining sublevel $3sp^3$ is empty, as indicated in **Figure 2a**.

Also, the nitrogen hybridization rises sublevels $2s^22p^3$, which yields four hybrids $2sp^3$ sublevels; one is a $2sp^3$ sublevel full of two electrons, and three sublevels $2sp^3$ are semi-complete with one electron each. Therefore, in the bonds between the atoms of aluminum and nitrogen, as shown in **Figure 2b**, there are three covalent bonds between three semi-complete sublevels hybrids of one aluminum atom (atom 4) with three atoms of nitrogen (atoms 1, 2, and 3), forming a tetrahedron of a length l_1 = 0.1885 nm and 110.5° angle. There is also one covalent bond between the empty hybrid sublevel of aluminum (atom 4) and the complete hybrid sublevel of nitrogen (atom 5), of ionic character, of length l_2 = 0.1917 nm, and angle of 107.7° between l_1 and l_2. The nitrogen atom (atom 5) joins with three additional aluminum atoms (atoms 6, 7, and 8) through covalent bonds between semi-complete hybrid sublevels, forming another tetrahedron of length l_1 and angle of 110.5°. The three aluminum atoms (atoms 6, 7, and 8) also bind

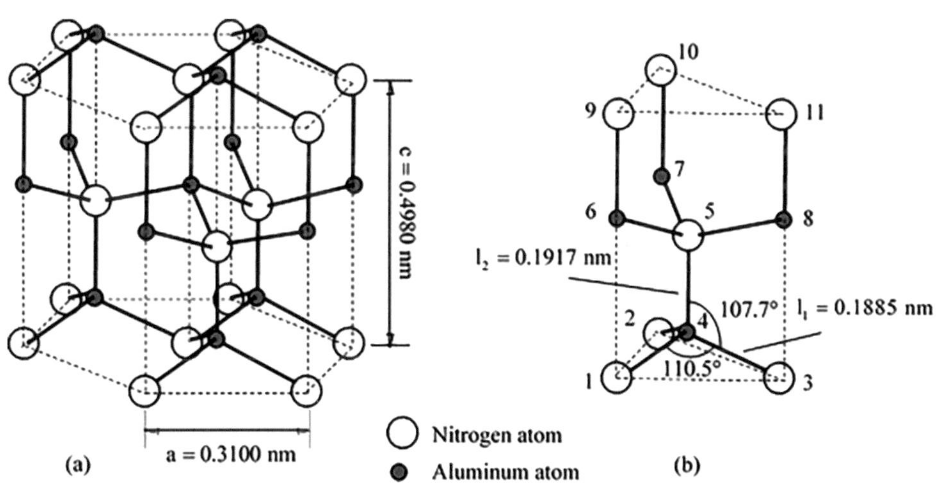

Figure 3.
(a) Bonds between aluminum and nitrogen atoms forming a hexagonal wurtzite AlN unit cell with the following lattice parameters a = b≅0.3100 nm and c = 0.4980 nm. (b) Bonds between atoms of aluminum and nitrogen form a prism of triangular base.

to other three nitrogen atoms (atoms 9, 10, and 11) through covalent bonds between the aluminum empty hybrid sublevel and the nitrogen complete hybrid sublevel, of ionic character, having a length of l_2 and 107.7° angle between l_1 and l_2. Those 11 atoms of aluminum and nitrogen (atoms 1, 2, 3, 4, 5, 6, 7, 8, 9, 10, 11) have the form of a prism of triangular base. This prismatic structure originates the wurtzite hexagonal AlN unit cell, as shown in **Figure 3a**, that presents the following lattice parameters a = b≅0.3100 nm and c = 0.4980 nm. The orientation of the crystal axes and of the planes for the hexagonal unit shown in **Figure 4a** are (100, 001), respectively.

The vibrational phonon modes of the AlN film, E_1(TO), and A_1(TO) active in the IR are associated with the respective covalent bonds l_1 and l_2 as indicated in **Figure 2b**. In addition, they absorb energy according to the crystal orientation of the AlN film and polarization of the electric field of the IR beam, so this feature can be used to identify the preferred orientation of the film. In the IR analysis, the electric field of the beam is polarized in parallel to surface of the thin film, that is, the incident electric field is perpendicular to the c axis. The ratio between the energy absorbed by the phonon modes E_1(TO) and A_1(TO) is defined by Eq. (1).

$$A_f \frac{\text{Energy absorbed } by \text{ A}_1\text{ (TO)}}{\text{Energy absorbed } by \text{ E}_1\text{ (TO)}}. \tag{1}$$

If the ratio A_f is much smaller than one, it indicates a high degree of AlN (001) crystalline orientation. On the other hand, A_f greater than one indicates a degree of AlN (100) crystal orientation. Thus, it can be concluded that when the phonon mode E_1(TO) absorbs more energy than mode A_1(TO), the crystalline orientation degree tends to AlN (001), whereas if phonon mode A_1(TO) absorbs very more energy that mode E_1(TO) the preferred orientation tends strongly to AlN (100).

Figure 5 shows, highlighted in red, the thin film growth units AlN (100) and AlN (001). The atoms and/or clusters of atoms, that reach the surface of the substrate and are adsorbed, move along the surface, colliding and combining, giving rise to nucleation,

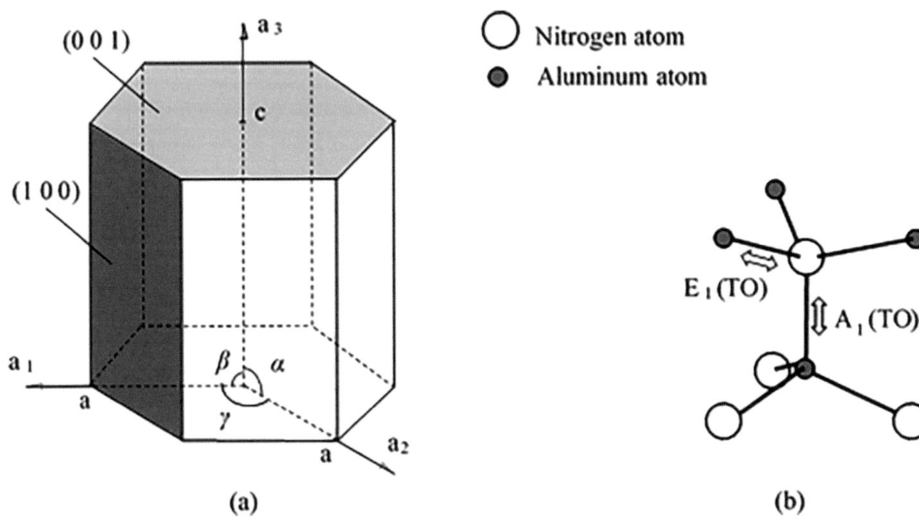

Figure 4.
(a) Hexagonal unit cell of the AlN (a = b≠c, α = β = 90°, and γ = 120°) with your crystal axis a_1, a_2, and a_3 plus the identification of the planes (100) and (001). (b) Vibrational phonon modes of the thin film AlN, actives in the IR, E_1(TO), and A_1(TO).

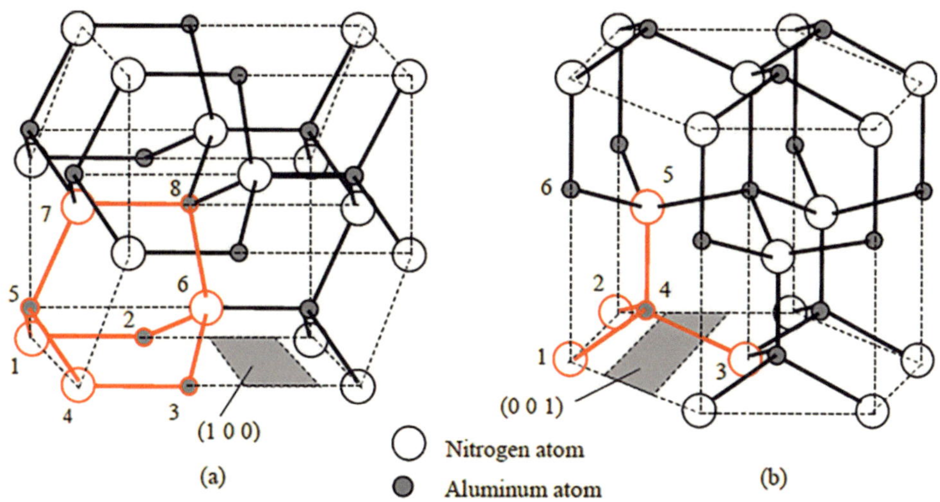

Figure 5.
Aluminum Nitride hexagonal unit cell growth units highlighted in red: (a) AlN (100) and (b) AlN (001).

which grows with the arrival of more species (atoms and/or clusters) forming larger nuclei (islands), that by coalescence end up forming a larger whole, until a continuous layer emerges, and this whole process repeats itself until a thin film is produced. A better explanation for the formation process of AlN (100)/Si (100) thin films, in which the c-axis is parallel to the substrate surface is: when the mean free path is much smaller than the target substrate distance, the collisions between Al and N species occur more often in the space between the target and the substrate, thus forming dimers many Al-N, which are deposited on the substrate and the AlN (100) preferential orientation is achieved [16]. When the species involved in the growth are atoms and ions, then during the growth of AlN (100), **Figure 4a**, there are four difficult depositions (atoms 5, 6, 7, and 8) and, in the case of AlN (001), **Figure 4b**, there are two difficult depositions (atoms 4 and 5); Consequently, this favors the growth of the AlN with the c-axis perpendicular to the surface of the substrate. If the species involved in the growth are dimers, Al-N, then in the formation of AlN (100), **Figure 4a**, we have two difficult depositions (dimers 5-7 and 8-6) and in the formation of AlN (001), **Figure 4b**, two difficult depositions (dimers 4-3 and 6-5) and also, since all bonds of the AlN (100) growth unit are difficult to break and in the growth AlN (001) unit only one ionic bond easy to break, the AlN film has a good chance of growing with the c-axis parallel to the surface of the substrate.

During the deposition of AlN films (reactive magnetron sputtering and RF source) the degree of crystal orientation of the AlN films, (100) and (001), is strongly influenced by the energy of the species involved in the process. The energy of species that reach the substrate can be controlled by adjusting some parameters during deposition (i.e., mean free path, temperature, pressure, target-substrate distance). In this scenario, the mean free path (L) and the collision probability (Q) of the species are given by [17]:

$$L = \frac{\kappa_B T}{\sqrt{2}(4\pi r_{Mean}^2)P} \tag{2}$$

and

$$Q = 1 - \exp(-D_{TS} / L) \tag{3}$$

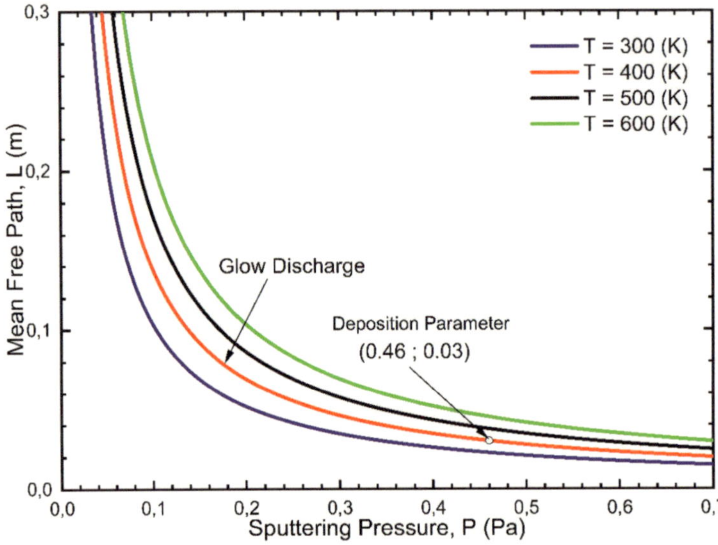

Figure 6.
Deposition parameters: L versus P. The curves were implemented for and typical plasma temperature values, and r_{Mean} = 150 nm. Reproduced from [17] with permission.

Where T is the absolute temperature, P is the plasma pressure, D_{TS} is the target-substrate distance, and r_{Mean} is the mean radius of the constituent species of the plasma, and $\kappa_B = 1.38 \times 10^{-23}$ J/K is the Boltzmann constant. **Figure 6** shows the estimation of the three deposition parameters: temperature, working pressure, and mean free path. For these deposition parameters, **Figure** 7 shows the high probability of obtaining crystalline and highly oriented AlN (100) thin films.

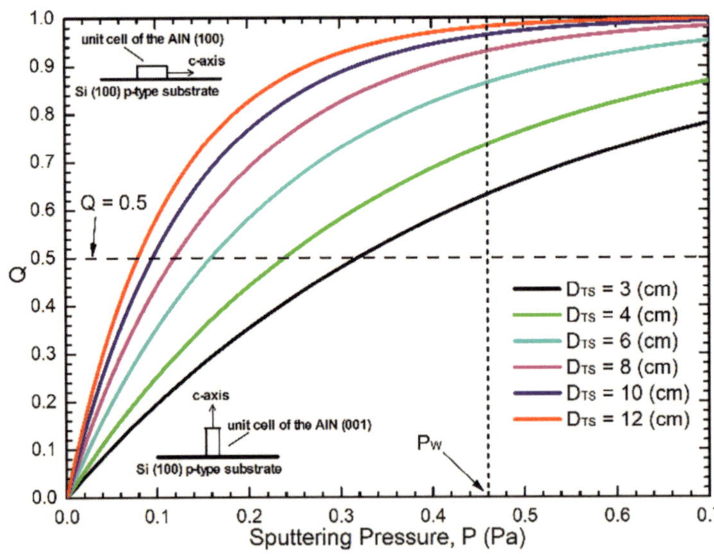

Figure 7.
Probability of obtaining crystalline and highly oriented AlN (100) thin films. Curves implemented for glow discharge, different values of D_{TS}, and r_{Mean} = 150 nm. Reproduced from [17] with permission.

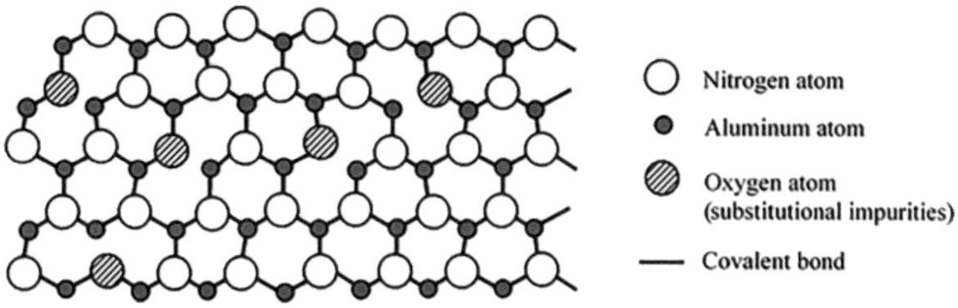

Figure 8.
The substitutional defect: layers of atoms in the formation of the wurtzite lattice of the AlN with the incorporation of some oxygen atoms in place of some nitrogen atoms.

The species energy can be increased by increasing the RF power or the substrate temperature, or decreasing the working pressure, or the distance between the target and the substrate. The Al and N species arrive with too much energy on the substrate surface, favoring the degree of crystal orientation AlN (001), where the (001) plane is parallel to the substrate surface. Now, the orientation degree of AlN (100) with (100) plane parallel to the substrate surface is strongly favored when the energy of Al-N species (dimer) is smaller, in which case many collisions occur prior to deposition.

3. Incorporation of oxygen into AlN crystal lattice

Generally, thin films are found to be defective due to the presence of impurities. A common defect in AlN thin films is the presence of a substitutional defect - N^{3-} ions of the AlN crystal lattice are replaced by O^{2-} ions - according to **Figure 8**.

This occurs because the characteristics (ionic radius and electronegativity) of oxygen and nitrogen are quite similar. The ionic radius of the oxygen is $r_O = 0.140$ nm and of the nitrogen $r_N = 0.146$ nm gives a difference of ionic radii $\Delta r = 4.11\%$. The electronegativity of oxygen is $\xi_O = 3.5$, of aluminum $\xi_{Al} = 1.6$, of nitrogen $\xi_N = 3.0$, and therefore the differences in electronegativity with respect to Al are $\Delta\xi_{O\text{-}Al} = 1.9$ (ionic covalent bonds) and $\Delta\xi_{N\text{-}Al} = 1.4$ (polar covalent bonds). The impurity atoms give rise to stress in the AlN (100) crystalline lattice since the ionic radius difference between N and O is very small ($\Delta r<15\%$) [18]. Otherwise, the impurity atoms will create substantial distortions in the crystalline lattice and may form a new phase. Also, during the process of thin film growth, there is a strong competition between oxygen and nitrogen because the high electropositivity of aluminum and high electronegativity of the oxygen favors the formation of an intermediate compound (Al_2O_3), instead of material with substitutional defect.

4. Growth of AlN films

4.1 Deposition techniques of AlN films

Aluminum nitride (AlN) films can be deposited on various substrate types via physical and chemical deposition methods as shown in **Figure 9**. Typical deposition methods include sputtering processes, metalorganic chemical vapor deposition (MOCVD), Molecular-beam epitaxy (MBE), and pulsed laser deposition (PLD).

Sputtering methods are the most used to grow AlN films due to their advantages such as low temperature, low cost, and flexibility. The sputtering deposition allows

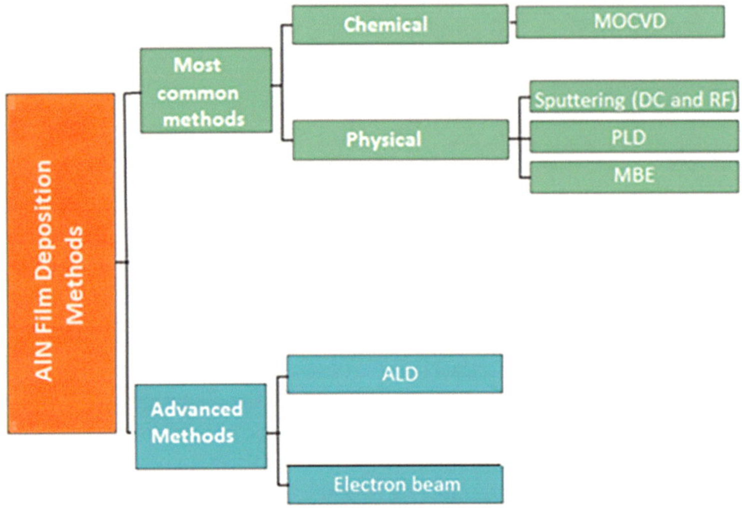

Figure 9.
Overview of the main methods for deposition of AlN films.

the development of devices and sensors on different types of substrates including polymeric substrates. Recently, Cunha et al. reported growth of highly (100)-oriented aluminum nitride (AlN) thin films on (100) Si substrate, in poor vacuum systems, by radio frequency magnetron sputtering (**Figure 10a**). High-quality films with good stoichiometry Al/N (≈1:1) and low oxygen concentration (<10%) were produced by varying the target-substrate distance and the deposition time, whereas the temperature, the nitrogen flow, RF power, and sputtering pressure were fixed [17].

Ghosh et al. discussed the growth of both the undoped and doped AlN films on GaN/Sapphire templates in an MBE chamber (**Figure 10b**) using a plasma-assisted process called PAMBE (plasma-assisted molecular beam epitaxy). The authors concluded that employing an excimer laser annealing with optimized power and frequency rather than the conventional thermal annealing can be a potential alternative route toward improving the structural and electrical properties of AlN layers [19].

An AlN interfacial passivation layer prepared in an MOCVD system (**Figure 10c**) was reported by Aoki et al. The proposed of this study was to present a route for the fabrication and optimization of GaAs metal–oxide–semiconductor (MOS) structures comprising an Al_2O_3 gate oxide, deposited via atomic layer deposition (ALD) and using an AlN interfacial passivation layer [20].

Pulsed-laser-deposited AlN films were produced by Vispute et al. using the system shown in **Figure 10d** at substrate temperatures ranging from 25°C (room temperature) to 1000°C. The AlN films were employed in the fabrication of device-quality AlN heterostructures grown on SiC for high-temperature electronic devices [21].

In relation to emerging advanced AlN deposition methods, atomic layer deposition (ALD) can be highlighted. Amorphous AlN films obtained by ALD were investigated by Parkhomenko et al. using trimethylaluminum and monomethylhydrazine as the precursors at a deposition temperature of 375–475°C. The ALD AlN films exhibited an oxygen content of as low as 4%. In addition, they were compact, continuous and with mechanical properties comparable to those of AlN films produced by other techniques [22]. ALD of AlN on different SiC surfaces with different crystallographic orientation was also investigated recently [23]. For all layers, the surface morphology

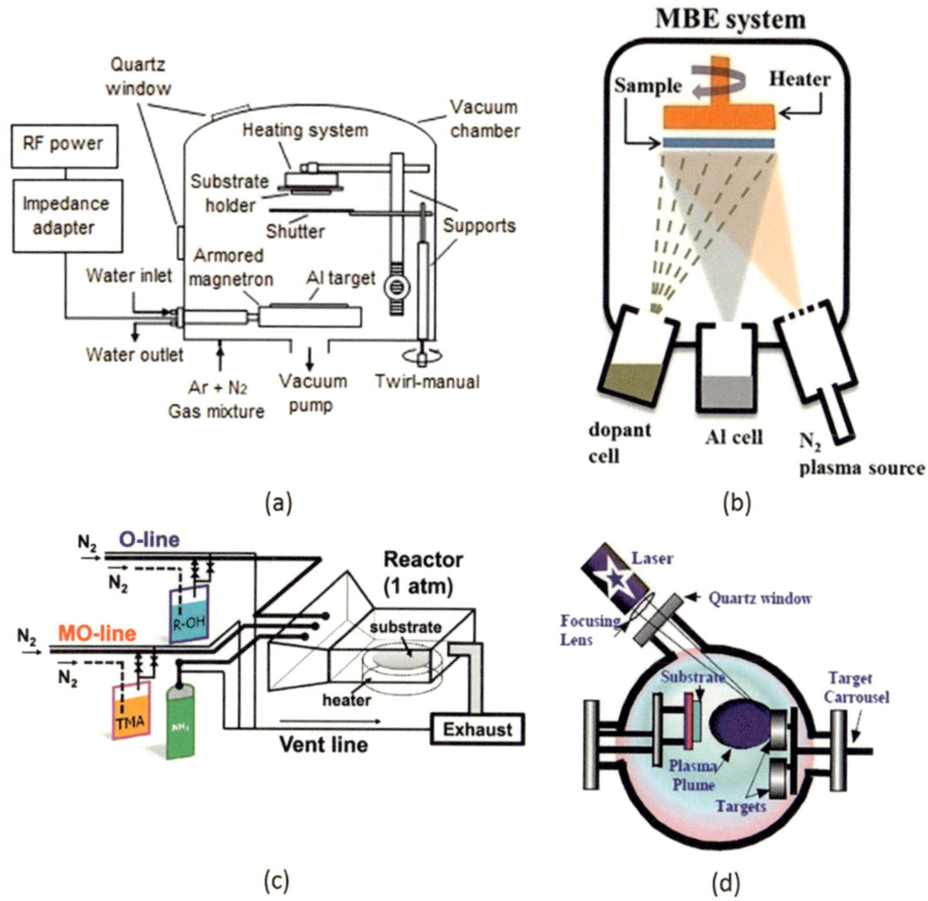

Figure 10.
Examples of systems used to grow AlN films: (a) sputtering [17], (b) MBE [19], (c) MOCVD [20] and (d) PLD [21].

and the chemical composition results showed that the ALD AlN films exhibit good characteristics films for surface acoustic wave (SAW) devices. The same authors also reported the morphological evolution of ALD AlN films on 4H-SiC substrates [24].

The growth of AlN films has been exploited on different substrates to form both a buffer layer and a sensing layer (**Figure 11**). The AlN buffer layer is used to improve the growth and properties of other thin films, such as GaN, SiC, ZnO, and diamond among others, for several applications, whereas AlN sensing layers are used mainly in piezoelectric devices, for example SAW (surface acoustic wave) sensors.

4.2 Growth of AlN buffer layer

Several studies have been devoted to the growth of AlN buffer layers to be used in different applications as those illustrated in **Figure 12**.

Zhang et al. discussed the growth of sputtered highly oriented AlN films on Si (100) and Si (111) substrates to use them as a proper buffer layer for epitaxial growth of gallium nitride (GaN) films. It was observed that the AlN (0001) films grown on Si (100) exhibit large strain due to the large lattice mismatch between these materials, whereas the AlN films grown on Si (111) have strain dependent on the discharge

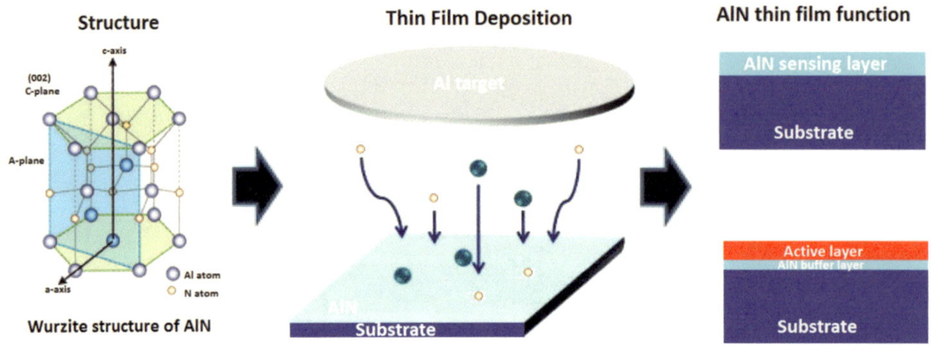

Figure 11.
Illustration of AlN structure, thin film deposition and thin film application as sensing layer and buffer layer.

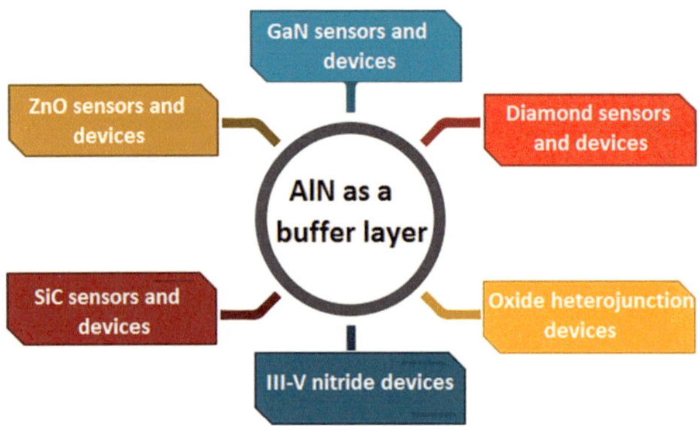

Figure 12.
Examples of applications of AlN buffer layers.

power in sputtering. Therefore, they concluded that the orientation of the Si substrates and the discharge power impact greatly the strain of sputtered AlN films [25].

The growth of sputtered AlN buffer layer on Si (111) was also reported by Núñez-Cascajero et al. In their paper, they addressed the use of AlN as a buffer layer for the development of AlInN/p-Si heterojunction solar cells. For this, it was investigated the influence of power applied to the Al target on the properties of AlN on Si (111). They found that the presence of the AlN buffer layer leads to an improvement of the structural quality of the $Al_{0.37}In_{0.63}N$ and that the solar cells based on this material show good rectifying behavior in the dark [26].

The growth of amorphous SiC thin films on AlN buffer layers deposited on glass and Si substrates was reported by Wang et al. It evaluated the effect of AlN buffer layer thickness on the morphological and mechanical properties of the SiC. Overall, their results indicated that the AlN buffer layer can effectively improve the adhesion strength of SiC thin films [27]. In another study focused on a comparison among substrates for the development of SiC thin film piezoresistive sensors, Fraga et al. evaluated the piezoresistive properties of SiC films grown on AlN/Si [28]. It evidenced the importance of growing SiC film on AlN in order to develop piezoresistive sensors for high-temperature applications.

In addition to silicon, other substrates are being used in the deposition of AlN buffer layer by sputtering. In a recent paper published in the Journal Materials Science in Semiconductor Processing, the effects of the use of sputtered AlN buffer layer on the carrier transport properties of p-NiO/n-InN heterojunction diode were investigated. In this study, AlN films were grown Al_2O_3 substrates with varying N_2 flow rates in the sputtering process. In order to analyze the reasons for the deterioration of the device characteristics, the influence of AlN buffer layer on I–V characteristics of the heterojunction diode was studied in the temperature range of 30–110°C. A good performance was observed for the heterojunction diode fabricated [29]. Chen et al. also investigated sputtered AlN films on sapphire substrates. However, their focus was to release the film stress using a post-deposition rapid thermal annealing (RTA) at 700–900°C for 5 min. The Raman spectra showed that the in-plane tensile stress of deposited AlN films is released by the RTA [30].

Regarding the use of AlN films as a buffer layer for diamond growth, Mandal et al. carried out the growth of thick (>100 μm) CVD diamond layers on AlN with low thermal boundary resistance between diamond and AlN. In their study, they used a metalorganic chemical vapor deposition (MOCVD) system to grow a 250-nm-thick AlN layer on 150 mm Si substrates. It was highlighted that diamond/AlN could be used for thermal management of GaN high-power devices [31]. Most recently, Forsberg et al., a high sensitivity infrared spectroscopy with a diamond waveguide on aluminum nitride [32].

Zinc oxide (ZnO) films have been also grown on AlN buffer layer for electronic device applications. A recent paper evaluated the temperature-dependent electrical transport properties of n-ZnO/AlN/p-Si heterojunction diodes [33]. Both AlN and ZnO films were deposited by RF magnetron sputtering. Results showed that the use of AlN buffer layer improved electrical and structural characteristics because the AlN between ZnO and Si lowers the mismatch in thermal expansion coefficient/lattice. In a previous study, Xiong et al. exploited the growth of ZnO films on a 150 nm AlN buffer layer on -sapphire substrates. It was noted that c-plane ZnO growth on c-plane sapphire by PLD at slight rough surface morphology of AlN buffer layer can result in a significant variation of ZnO crystallinity [34].

4.3 Growth of AlN active sensing layer

The excellent piezoelectric properties of AlN films have motivated studies on the development of sensors based on these materials. Tonisch et al. reported the piezoelectric properties of polycrystalline AlN thin films on Si (111) substrates for MEMS applications [35]. Reactive dc-sputtering and metalorganic chemical vapor deposition (MOCVD) were used to deposit the AlN films. The piezoelectric coefficient d_{33eff} of AlN thin films was measured using two techniques: piezoresponse force microscopy and an interferometric technique. The value of the effective piezoelectric coefficient d_{33} for the prepared AlN thin films remained as high as 5.1 pm/V even for lower degrees of texture.

In the same year, the temperature dependence of the piezoelectric coefficient d_{33} of sputtered AlN film on a polycrystalline silicon/silicon dioxide/silicon wafer measured at temperatures up to 300°C was reported by Kano et al. It was observed that the piezoelectric coefficient d_{33} has a constant value at temperatures ranging between 20°C and 300°C [36].

A more recent paper shows that the high (002) orientation AlN films have uniform piezoelectric performance [37]. In this study, AlN thin films were grown on Pt/Ti/SiO_2/Si (100) substrates by an optimized magnetron sputtering process. A high-frequency SAW device (f_o = 4.47 GHz) was constructed based on optimal AlN films. In another study on optimization of AlN growth, Cao et al. optimized DC magnetron sputtering, the process by controlling the distance target-substrate in the deposition of AlN thin films deposited on Si and Pt substrates [38]. A (002) AlN preferred

Device type	AlN film deposition method	Substrate	AlN film orientation	AlN film thickness (nm)	Piezoelectric coeficient (pm/V)
Implantable biomedical MENS devices	Sputtering	Pt/Ti/ UNCD	(002)	200	5.3
Flexible SAW sensors	Middle-frequency magnetron sputtering	Y_2O_3/ Hastelloy	(002)	3000	5.02
Sensory devices	Magnetron sputtering	(100) Silicon	(002)	2000	5.57
SMA ME sensor	DC reactive sputtering	TiNiCu shape memory alloys	(002)	2000	6.3
Piezoelectric accelerometers	DC sputtering	Pt/Ti/SiO_2/ Si	(002)	1000	5.92
High-frequency flexible SAW sensors	DC reactive magnetron sputtering	Mo/ Polymide	(002)	800	8.01
Biocompatible piezo-optrodes devices	DC sputtering	Mo/Optical fiber	(002)	1000	5.4

Table 1.
Piezoelectric device types based on sputtered AlN films. Adapted from [42].

orientation was obtained with D_{TS} = 4 cm. They concluded that under optimum conditions, the as-deposited AlN films show uniform piezoelectric properties and favorable read and write performance [38].

The piezoelectric properties of sputtered Sc-doped AIN polycrystalline films on 200 mm Si wafers were also evaluated in order to use them as active layers for high frequency (GHz range) acoustic resonators [39]. The piezoelectric activity of the as-deposited AlScN films were improved after a 15 min post-deposition annealing at 600°C, leading to a 20% increase in the electromechanical coupling factor [39].

The influence of He implantation on piezoelectric properties of epitaxial AlN thin films were discussed recently [40]. It was noted that while He implantation induces uniaxial strain, it decreases d_{33} due to implantation-induced N site disorder [40].

Pressure gradient sputtering (PGS) has also been used to grow AlN films. The piezoelectric constant (d_{33}) of the AlN grown by the PGS method was higher than that of the conventional method indicating which the PGS technique has an advantage in the growth of AlN films with highly c-axis oriented and a single dielectric domain [41].

In September 2021, an open research knowledge graph (ORKG) comparison devoted to aluminum nitride films and their applications in piezoelectric devices was published [42]. **Table 1** summarizes the characteristics of some piezoelectric devices based on AlN films.

5. SAW sensors based on AlN films

In recent years, much progress has been carried out on surface acoustic wave (SAW) sensors and applications have been reported in the fields of microfluidics, chemical, biomedical, and mechanical as illustrated in **Figure 13**.

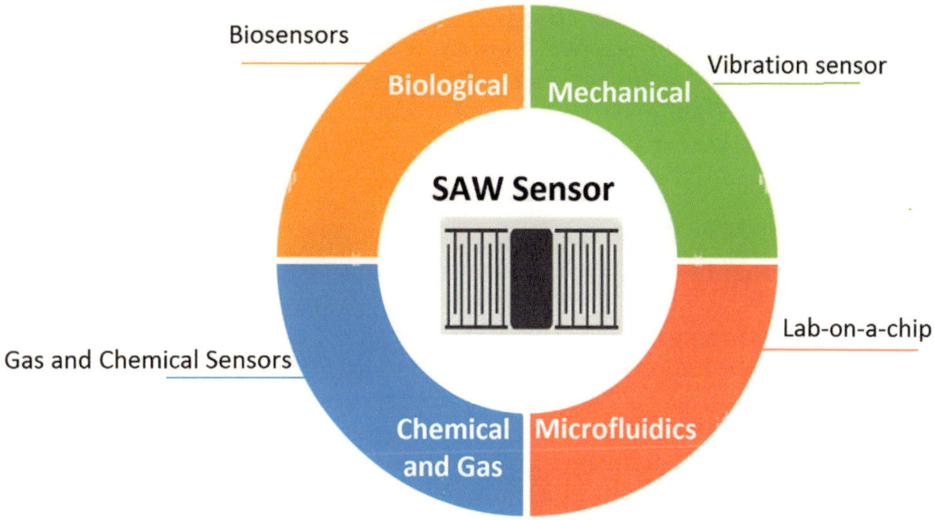

Figure 13.
Some relevant applications of SAW sensors.

In the literature, various AlN-based sensors are reported. In 1991, Odintzov MA et al. evaluated sputtered AlN films for SAW sensors. AlN films were grown on glass substrates by reactive RF magnetron sputtering and their piezoelectric properties were analyzed. The obtained film was used to implement a SAW temperature sensor. Moreover, it discussed the importance of obtaining polycrystalline AlN-oriented films with perfect crystallization [43].

A SAW-based sensor pressure with a sensibility of 0.33 MHz/bar based on AlN deposited on free-standing diamond substrates was developed by Rodríguez-Madrid. The influence of the piezoelectric film thickness on the SAW response was evaluated. Optimized AlN thin films of 300 nm were used to fabricate one-port SAW resonators operating in the 10–14 GHz frequency range, which were used as SAW pressure sensors [44].

A 3 μm-thick AlN/Sapphire-based SAW resonators with high-quality factors for high-temperature applications (up to 600°C) were fabricated by Streque et al. The quasi-synchronous resonators proposed remained well-tuned for temperatures up to 400°C, and show high-quality factors, as high as 3400 at 400°C [45].

A new two-step growth process integrating metalorganic chemical vapor deposition (MOCVD) and physical vapor deposition (PVD) technologies were proposed to grow AlN films on Si substrates by Xinyan et al. [46]. High-quality AlN-based FBARs wafers were obtained, showing that the Q-factor of FBARs with two-step grown AlN precedes FBARs with one-step grown PVD AlN by 57.6% [46].

6. Conclusion

As shown in this book chapter, the excellent physical, chemical, dielectric, thermal and mechanical properties of aluminum nitride (AlN) thin films have stimulated great interest in these materials and their wide applications. Until today, sputtering processes are the most used to grow AlN films. In recent years, the ALD technique, an outstanding process in nanotechnology, has been explored to grow AlN films. This book chapter also presented various applications for AlN films as a buffer layer and as a sensing layer. Among the applications, the use of AlN sensing layers is the most discussed in the literature due to their excellent piezoelectric properties and the good performance of SAW sensors based on these materials. In summary, this book chapter presents the key aspects of aluminum nitride thin film technology and its applications in electronic devices and sensors.

Acknowledgements

This work was supported by the Federal University of Itajubá (UNIFEI), Coordination for the Improvement of Higher Education Personnel (CAPES), National Council of Technological and Development (CNPq), and Minas Gerais State Research Support Foundation (FAPEMIG).

Conflict of interest

The authors declare no conflict of interest.

Author details

Cícero L.A. Cunha[1*], Tales C. Pimenta[1] and Mariana Amorim Fraga[2,3]

1 Federal University of Itajubá, Itajubá, Brazil

2 Federal University of São Paulo, São José dos Campos, SP, Brazil

3 Mackenzie Presbyterian University, São Paulo, SP, Brazil

*Address all correspondence to: cicerocunha@unifei.edu.br

References

[1] Berg NG, Paskova T, Ivanisevic A. Tuning the biocompatibility of aluminum nitride. Materials Letters. 2017;**189**:1-4. DOI: 10.1016/j.matlet.2016.11.041

[2] Fu S, Li Q, Gao S, Wang G, Zeng F, Pan F. Quality-enhanced AlN epitaxial films grown on c-sapphire using ZnO buffer layer for SAW applications. Applied Surface Science. 2017;**402**:392-399. DOI: 10.1016/j.apsusc.2017.01.025

[3] Fu YQ, Luo JK, Nguyen NT, Walton AJ, Flewitt AJ, Zu XT, et al. Advances in piezoelectric thin films for acoustic biosensors, acoustofluidics and lab-on-chip applications. Progress in Materials Science. 2017;**89**:31-91. DOI: 10.1016/j.pmatsci.2017.04.006

[4] Melo-Máximo L, Lin J, Murillo AE, Sales O, Oliva-Ramírez J, Oseguera J, et al. Deposition of AlN films for acoustic biosensors by deep oscillation magnetron sputtering: Effect of bias voltage. Thin Solid Films. 2018;**664**:83-89. DOI: 10.1016/j.tsf.2018.08.022

[5] Murillo AE, Melo-Máximo L, García-Farrera B, Martínez OS, Melo-Máximo DV, Oliva-Ramírez J, et al. Development of AlN thin films for breast cancer acoustic biosensors. Journal of Materials Research and Technology. 2019;**8**(1):350-358. DOI: 10.1016/j.jmrt.2018.02.007

[6] Prasad M, Kumar R. Deposition and process development of AlN for MEMS acoustic sensor. Vacuum. 2018;**157**:349-353. DOI: 10.1016/j.vacuum.2018.08.062

[7] Reusch M, Cherneva S, Lu Y, Žukauskaitė A, Holc K, Datcheva M, et al. Microstructure and mechanical properties of stress-tailored piezoelectric AlN thin films for electro-acoustic devices. Applied Surface Science. 2017;**407**:307-314. DOI: 10.1016/j.apsusc.2017.02.147

[8] Lin J, Chistyakov R. C-axis orientated AlN films deposited using deep oscillation magnetron sputtering. Applied Surface Science. 2017;**396**:129-137. DOI: 10.1016/j.apsusc.2016.11.025

[9] Ababneh A, Schmid U, Hernando J, Sánchez-Rojas JL, Seidel H. The influence of sputter deposition parameters on piezoelectric and mechanical properties of AlN thin films. Materials Science and Engineering B. 2010;**172**(3):253-258. DOI: 10.1016/j.mseb.2010.05.026

[10] Mele A, Giardini A, Di Palma TM, Flamini C, Okabe H, Teghil R. Preparation of the group III nitride thin films AlN, GaN, InN by direct and reactive pulsed laser ablation. International Journal of Photoenergy. 2001;**3**(3):111-121. DOI: 10.1155/S1110662X01000137

[11] Mishra M, Krishna S, Aggarwal N, Gupta G. Influence of metallic surface states on electron affinity of epitaxial AlN films. Applied Surface Science. 2017;**407**:255-259. DOI: 10.1016/j.apsusc.2017.02.128

[12] Riah B, Ayad A, Camus J, Rammal M, Boukari F, Chekour L, et al. Textured hexagonal and cubic phases of AlN films deposited on Si (100) by DC magnetron sputtering and high power impulse magnetron sputtering. Thin Solid Films. 2018;**655**:34-40. DOI: 10.1016/j.tsf.2018.03.076

[13] Dimitrova V, Manova D, Paskova T, Uzunov T, Ivanov N, Dechev D. Aluminum nitride thin films

deposited by DC reactive magnetron sputtering. Vacuum. 1998;**51**(2):161-164. DOI: 10.1016/S0042-207X(98)00150-X

[14] Molleja JG, Gómez BJ, Ferrón J, Gautron E, Bürgi J, Abdallah B, et al. AlN thin films deposited by DC reactive magnetron sputtering: Effect of oxygen on film growth. European Physical Journaal. 2013;**64**(2):20302

[15] Signore MA, Taurino A, Valerini D, Rizzo A, Farella I, Catalano M, et al. Role of oxygen contaminant on the physical properties of sputtered AlN thin films. Journal of Alloys and Compound. 2015;**649**:1267-1272. DOI: 10.1016/j.jallcom.2015.05.289

[16] Xu X-H, Wu H-S, Zhang C-J, Jin Z-H. Morphological properties of AlN piezoelectric thin films deposited by DC reactive magnetron sputtering. Thin Solid Films. 2001;**388**(1-2):62-67. DOI: 10.1016/S0040-6090(00)01914-3

[17] Cunha CLA, Pimenta TC, Fraga MA. Growth and properties of sputtered highly (100)-oriented oxygenated AlN thin films for SAW sensing applications. Microsystem Technologies. 2021;**27**:3773-3782. DOI: 10.1007/s00542-020-05165-1

[18] Callister WD Jr, Rethwisch DG. Materials Science and Engineering: An Introduction. 8th ed. USA: John Wiley & Sons; 2010

[19] Kankat G, Pratik B, Sudipta D. Excimer laser annealing: An alternative route and its optimization to effectively activate Si dopants in AlN films grown by plasma assisted molecular beam epitaxy. Materials Research Bulletin. 2018;**97**:300-305

[20] Takeshi A, Noboru F, Takenori O, Hiroyuki S, Masahiko H, Takayuki I. Electrical properties of GaAs metal-oxide-semiconductor structure Comprising Al_2O_3 gate oxide and AlN passivation layer fabricated in situ using a metal–organic vapor deposition/atomic layer deposition hybrid system. AIP Advances. 2015;**5**(8):087149. DOI: 10.1063/1.4929371

[21] Vispute RD, Patel A, Baynes K, et al. Pulsed-laser-deposited AlN films for high-temperature SiC MIS devices. MRS Internet Journal of Nitride Semiconductor Research. 2000;**5**:591-597. DOI: 10.1557/S1092578300004804

[22] Parkhomenko Roman G. Amorphous AlN films grown by ALD from trimethylaluminum and monomethylhydrazine. Dalton Transactions. 2021;**50**(42):15062-15070. DOI: 10.1039/d1dt02529e

[23] Beshkova M, Deminskyi P, Hsu C-W, Shtepliuk I, Avramova I, Yakimova R, et al. Atomic layer deposition of AlN on different SiC surfaces. Journal of Physics Conference Series. 2022;**2240**(1):012004. DOI: 10.1088/1742-6596/2240/1/012004

[24] Beshkova M, Blagoev BS, Mehandzhiev V, Yakimova R, Georgieva B, Avramova I, et al. Morphological evolution of thin AlN films grown by atomic layer deposition. Journal of Physics Conference Series. 2022;**2240**(1):012005. DOI: 10.1088/1742-6596/2240/1/012005

[25] Zhang JX, Cheng H, Chen YZ, Uddin A. Growth of AlN films on Si (100) and Si (111) substrates by reactive magnetron sputtering. Surface and Coating Technology. 2005;**198**(1-3):68-73. DOI: 10.1016/j.surfcoat.2004.10.075

[26] Núñez-Cascajero A, Valdueza-Felip S, Blasco R, de la Mata M, Molina SI, González-Herráez M, et al. Quality improvement of AlInN/p-Si heterojunctions with AlN buffer layer deposited by RF-sputtering. Journal of Alloys and Compounds.

2018;**769**:824-830. DOI: 10.1016/j.jallcom.2018.08.059

[27] Wang Y, Zha-ma T, Zhenming Y, Hui S, Jianhong G, Gao J. Research on adhesion strength and optical properties of SiC films obtained via RF magnetron sputtering. Chinese Journal of Physics. 2020;**64**:79-86. DOI: 10.1016/j.cjph.2020.01.006

[28] Fraga Mariana A, Humber F, Rasia Luiz A, Koberstein Leandro L. Effects of the substrate on piezoresistive properties of silicon carbide thin films. ECS Transactions. 2012;**44**(1):1375-1380. DOI: 10.1149/1.3694474

[29] WenBo P, YiJian Z, GuoJiao X, Yue L, JiaHui Z, JinMing Z, et al. Preparation of AlN thin film and the impacts of AlN buffer layer on the carrier transport properties of p-NiO/n-InN heterojunction by magnetron sputtering. Materials Science in Semiconductor Processing. 2022;**141**:106417

[30] Yanpu C, Zhu Z, Haolei Z, Ma P, Liren L, Zhu W, et al. Stress relaxation of three dimensional textured AlN films on sapphire substrate by rapid thermal annealing. Diamond and Related Materials. 2021;**118**:108532

[31] Soumen M, Chao Y, Fabien M, Pomeroy James W, Jerome C, Henry B, et al. Thick, adherent diamond films on AlN with low thermal barrier resistance. ACS Applied Materials & Interfaces. 2019;**11**(43):40826-40834. DOI: 10.1021/acsami.9b13869

[32] Pontus F, Patrik H, Mikael K. High sensitivity infrared spectroscopy with a diamond waveguide on aluminium nitride. The Analyst. 2021;**146**(22): 6981-6989. DOI: 10.1039/d1an01009c

[33] Prakash GC, Kumar SA, Jain Praveen K, Kant SS, Shilpi B, Sandeep S. Electrical transport properties of thermally stable n-ZnO/AlN/p-Si diode grown using RF sputtering. Materials Science in Semiconductor Processing. 2021;**128**:105734

[34] Xiong H, Dai JN, Hui X, Fang YY, Tian W, Fu DX, et al. Growth and properties of ZnO film grown on AlN buffer layer by PLD. In: International Photonics and Optoelectronics Meetings. 2012

[35] Tonisch K, Cimalla V, Foerster C, Romanus H, Ambacher O, Dontsov D. Piezoelectric properties of polycrystalline AlN thin films for MEMS application. Sensors and Actuators, A: Physical. 2006;**132**(2):658-663. DOI: 10.1016/j.sna.2006.03.001

[36] Kazuhiko K, Kazuki A, Yukihiro T, Morito A, Naohiro U, Nobuaki K. Temperature dependence of piezoelectric properties of sputtered AlN on silicon substrate. Sensors and Actuators, A: Physical. 2006;**130-131**:397-402. DOI: 10.1016/j.sna.2005.12.047

[37] Wang F, Fuliang X, Dianyou S, Lirong Q, Yulin F, Bangran F, et al. Research of micro area piezoelectric properties of AlN films and fabrication of high frequency SAW devices. Microelectronic Engineering. 2018; **199**:63-68. DOI: 10.1016/j.mee.2018.07.016

[38] Rongrong C, Wang F, Yulin F, Bangran F, Yinping M, Yang B, et al. Fabrication condition optimization of AIN films and its nanometer scale piezoelectric properties. Science of Advanced Materials. 2018;**10**(3):379-382. DOI: 10.1166/sam.2018.2988

[39] Marta C, Olivares Valerity FJ, Teona M, Jimena O, Enrique I. Effects of post-deposition vacuum annealing

on the piezoelectric properties of AlScN thin films sputtered on 200 Mm production wafers. In: IEEE International Ultrasonics Symposium. 2018

[40] Sharma V, Natali F, Kennedy J, Leveneur J, Fiedler H, Murmu P, et al. The effect of low energy helium implantation on the structural, vibrational, and piezoelectric properties of AlN thin films. Physica B: Condensed Matter. 2021;**601**:412481. DOI: 10.1016/j.physb.2020.412481

[41] Yoshikazu T, Kengo H, Ryo I, Hiroki O, Ken Y. Structural and piezoelectric properties of AlN thin films grown by pressure gradient sputtering. Japanese Journal of Applied Physics. 2022

[42] ORKG comparison. Available from: https://www.orkg.org/orkg/comparison/R141906/ [Accessed: September, 2021]

[43] Odintzov MA, Sushentzov NI, Kudryavtzev TL. AlN films for SAW sensors. Sensors and Actuators, A: Physical. 1991;**28**(3):203-206

[44] Rodríguez-Madrid JG, Iriarte GF, Williams OA, Calle F. High precision pressure sensors based on SAW devices in the GHz range. Sensors and Actuators, A: Physical. 2013;**189**:364-369. DOI: 10.1016/j.sna.2012.09.012

[45] Jérémy S, Julien C, Thierry L, Sami H-A, Hamid M'J, Mohammad R, et al. Design and characterization of high-Q SAW resonators based on the AlN/Sapphire structure intended for high-temperature wireless sensor applications. IEEE Sensors Journal. 2020;**20**(13):6985-6991. DOI: 10.1109/jsen.2020.2978179

[46] Xinyan Y, Lishuai Z, Peidong O, Hongbin L, Tielin Z, Guoqiang L. High-quality film bulk acoustic resonators fabricated on AlN films grown by a new two-step method. IEEE Electron Device Letters. 2022;**43**(6): 942-945. DOI: 10.1109/led.2022.3164972

Chapter 10

Thin-Film Batteries: Fundamental and Applications

Macdenis Egbuhuzor, Solomon Nwafor, Chima Umunnakwe and Sochima Egoigwe

Abstract

Thin-film batteries are solid-state batteries comprising the anode, the cathode, the electrolyte and the separator. They are nano-millimeter-sized batteries made of solid electrodes and solid electrolytes. The need for lightweight, higher energy density and long-lasting batteries has made research in this area inevitable. This battery finds application in consumer electronics, wireless sensors, smart cards medical devices, memory backup power, energy storage for solar cells, etc. This chapter discussed different types of thin-film battery technology, fundamentals and deposition processes. Also discussed in this chapter include the mechanism of thin-film batteries, their operation and the advantages of thin-film batteries over other batteries. The vast applications of thin-film batteries drive research in this area. These applications are discussed extensively.

Keywords: batteries, deposition, electrodes, seperator, solid state, substrate, thin films

1. Introduction

Thin-film batteries are nano- to millimeter-sized solid-state batteries comprising the anode, the cathode, the electrolyte and the separator. The anode is the negative electrode that is oxidized after giving up electrons to the external circuit. It is the anode that generates ions that move through the electrolyte. The cathode is the positive electrode that accepts electrons from the external circuit and is reduced in the process. During the charging and discharging process, ions are inserted into and extracted from the cathode. The electrolyte is the medium for charge transfer between the cathode and the anode. Thin-film electrolyte is usually chemically stable, ionically conductive and electrically insulating and is required also to build good contact with the cathode and anode surfaces. The separator prevents physical contact between the anode and the cathode without blocking the transport of ions. Most times in thin-film batteries, the solid electrolyte acts both as an ion transport medium and physically separates the cathode and the anode as shown in **Figure 1**.

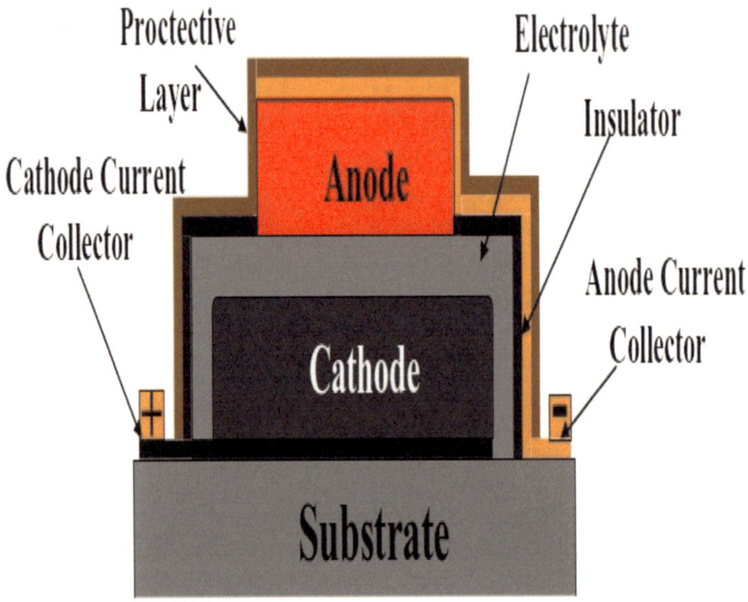

Figure 1.
Thin-film batteries with a solid-state cathode, anode, and electrolyte.

Thin-film batteries are manufactured using physical and chemical deposition techniques [1]. They include magnetron sputtering, pulsed laser deposition, molecular layer deposition [2], atomic layer deposition, vacuum evaporation [3], thermal evaporation, electron beam and sputtering [4]. These techniques follow four pathways called thin-film battery technologies [5].

The mechanism of the thin-film batteries is that ions migrate from the cathode to the anode charging and storing absorbed energy and migrating back to the cathode from the anode during discharge and thereby releasing energy [6].

The recent research in and development of smarter societies have necessitated the integration of smart devices with improved safety, specific energy, power and reduced-size materials [3]. This has given rise to the demand for using thin-film rechargeable batteries for electrical energy storage with good energy and power densities, excellent mechanical strength, good and long cycle life and appreciable temperature tolerance for small portable consumer electronics, especially in cell phones, laptops and notebook computers, smart cards, mobile applications, for electric cars, communication and other electrical equipment [7].

This book chapter reviews the fundamentals of thin-film batteries and the use of these batteries in various applications.

2. Types of thin-film battery technologies

Four major thin-film battery technologies are discussed here. They include (a) printed battery technology, (b) ceramic battery technology, (c) lithium polymer battery technology and (d) nickel-metal hydride (NiMH) button battery technology [5]. The choice of manufacturing of any of the technological paths will depend on the end-use application of the batteries.

2.1 Printed battery technology

Printed batteries can be in the form of rechargeable and non-rechargeable printed batteries. These printed batteries can be classified based on the make of the electrochemical systems as rechargeable and non-rechargeable printed batteries as shown in **Figure 2** [8]. The printing paste for batteries is made up of the electroactive material, the solvent, the conductive agent and the binder. The advantages of printed batteries include (1) it is possible to develop more than one layer pattern of design in the battery fabrication [9, 10], (2) Low cost and mechanically flexible batteries [11], (3) Thin and flexible energy storage [12] and (4) compatibility with the printing processes and devices [13]. Printed battery technologies find applications in the Internet of Things (IoT) devices [5], wearable electronic devices, smart cards, remote sensors, medical devices [9], microelectromechanical systems (MEMS) devices, low-power microprocessors, big-data analytics [14], electric vehicles, stationary storage grids [15], renewable energy and smart grid [16], etc.

2.2 Ceramic battery technology

They are all solid-state batteries made up of ceramic materials as their electrolytes. The demand for high-density batteries has necessitated research into more technologies that will enhance the thermal stability, mechanical strength and wettability of solid-state battery manufacture. Enhancements at the metal electrodes and/or separator boundaries are made possible by the introduction of ceramic batteries to boast some of the desired properties needed for solid-state batteries. Ceramic-based flexible sheet electrolytes have been formed to improve the energy density of solid-state batteries by synthesizing flexible composite Al-doped LLZO sheet electrolyte [17], coating of Al_2O_3, SiO_2 and TiO_2 onto polyethylene membrane seperators [18]. They

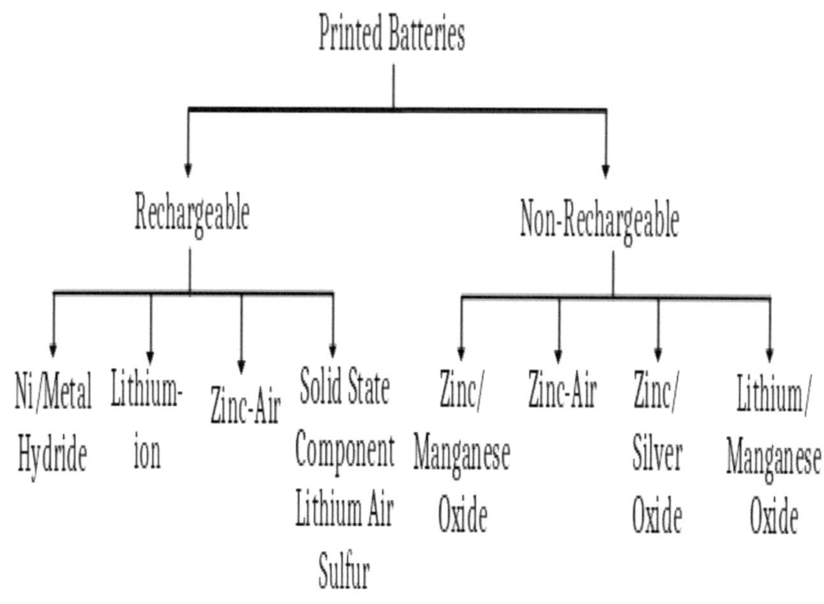

Figure 2.
Classification of printed battery technology.

are used in cogeneration and fuel cell, electric vehicles, mobile devices and stationary storage applications [19].

2.3 Lithium polymer battery technology

This works on the principle of intercalation and de-intercalation of lithium ions from the positive electrode (cathode) to the negative electrode (anode) and vice versa through a solid-state electrolyte that provides the conductive medium. Lithium-polymer batteries are lightweight, long-lasting and powerful solid-state batteries that can guarantee a constant energy supply [20]. They have a high energy density, flat voltage curves, low self-discharge and no memory effect from being discharged and charged again. The technology is used in laptop computers, personal electronics, cellular phones [21], notebooks and digital cameras. The polymer electrolyte may be in form of dry solid polymer electrolyte, polymer in-salt system, single-lithium-ion conducting electrolyte and gel polymer electrolyte [22].

2.4 Nickel-metal hydride (NiMH) button battery technology

This technology is an energy storage system that depends on charge/discharge reactions occurring between the nickel oxide-hydroxide cathode as the active material and the hydrogen-absorbing alloy anode [23]. This battery technology exhibits good tolerance to overcharge/discharge, high power capacity, very safe and compatible [24]. They are mainly used as power sources for hybrid electric vehicles, digital cameras and cell phones [25].

3. Deposition techniques

Thin-film deposition techniques are used to modify the surface properties of solid-state thin-film batteries. The modification affects the battery characteristics such as energy density, conductivity, storage capacity, charging and discharge time, etc.

Thin-film battery deposition techniques are classified into physical and chemical deposition processes as shown in **Figure 3** [26]. In the chemical deposition process, a chemical reaction takes place before the product is deposited on the substrate while the physical deposition method involves only physical mixing and deposition of the mixture on the substrate without any chemical reaction [27]. The coating material in a physical deposition is always a solid while the coating materials in chemical deposition are always in gaseous form. Physical deposition techniques include thermal evaporation, sputtering, ion plating and arc vapor deposition while chemical deposition techniques include chemical vapor deposition, plating, sol-gel deposition, chemical bath deposition and spray pyrolysis deposition techniques [28].

3.1 Physical deposition method

This method includes thermal evaporation, molecular beam epitaxy, pulsed laser deposition, ion plating evaporation, cathodic arc deposition and sputtering techniques. In thermal evaporation, molten metals or metal oxides form vapor and are deposited on the substrate as it cools [29, 30]. Molecular beam epitaxy (MBE) is a deposition technique in which a single crystal layer is deposited in a single crystal substrate, using molecular beams in an ultra-high vacuum chamber [31]. It is used in

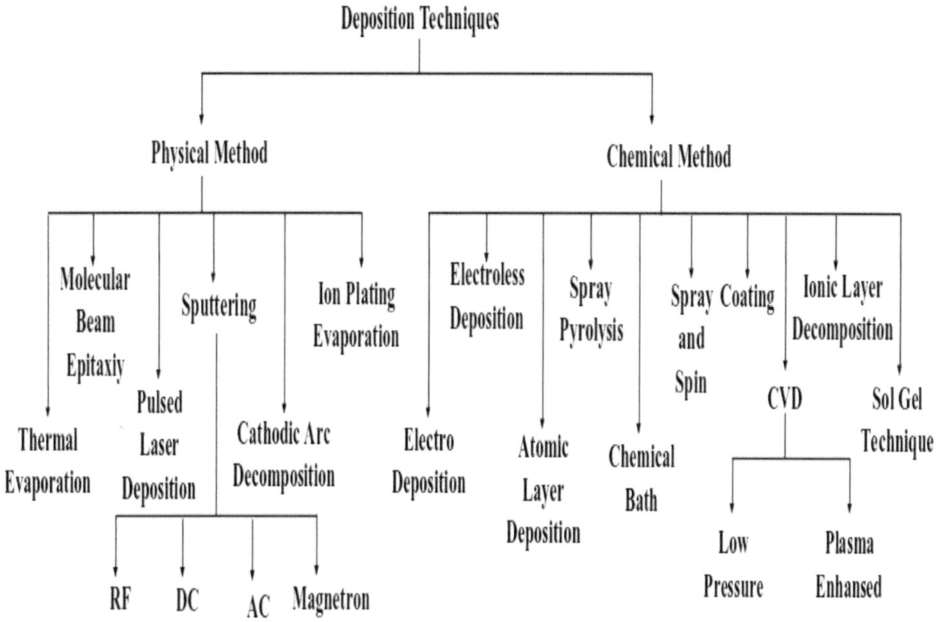

Figure 3.
Deposition techniques.

multi-junction solar cell applications [32], Ga-FACE GaN electron devices [33], etc. Pulsed laser deposition (PLD) process is a technique where a high-power pulsed laser beam strikes a target, vaporizes it and deposits it as a thin film on a substrate [34]. It is useful for optical applications, especially for UV laser emission [35] and superconductor devices for electronic and medical applications [36].

3.2 Chemical deposition method

Chemical deposition of thin films involves the generation of the atoms, molecules or ions through a chemical means; transportation of the atoms or ions through a medium and condensation and cooling of the atoms, molecules or ions on the substrate [27]. Chemical methods include chemical vapor deposition (CVD), electrodeposition, electroless deposition, spray pyrolysis, ionic layer deposition, sol-gel technique, chemical bath, spray and spin coating as shown in **Figure 2**.

Spray pyrolysis is the process of depositing a thin film on a heated surface by spraying a solution on the heated substrate [37]. The solution reacts with the heated substrate to form a thin film on the surface of the substrate. Mostly used for manufacturing semi-conductor alloys and conductive glasses. Electrodeposition is a technique of coating a thin layer of metal on top of a different metal through electrolysis by reducing the cations of the material from the electrolyte and depositing it as a thin film on the substrate [38]. It is a major method for the production of rechargeable lithium-ion batteries [39, 40]. Electroless deposition is a thin-film coating technique without the application of external electric power. It is used for non-conducting substrates and its major application is in the metalizing of printed wiring boards [41]. Atomic layer deposition (ALD) is a chemical deposition technique where chemical precursors introduced on the surface of the substrate react to form ultra-thin film monolayers on the substrate surface. They find application in fuel cells, capacitors, microelectronics and areas where highly uniform ultra-fine mono-layered thin films are required [42]. Sol-gel is a technique where metal alkoxides, used as precursors, are dissolved in a solvent (especially water or alcohol), heated and stirred to form a gel, which is condensed, dried and deposited on the substrate as a thin film [43]. The process involves hydrolysis, polycondensation, gelation, drying and crystallization [44].

3.3 Deposition technique for cathode materials in thin-film batteries

A cathode is a very important component of thin-film batteries. This electrode material facilitates the stability of the electrochemical reactions in the electrode/electrolyte interface. The material composition of the cathode determines the thermal stability and rate capacity of the battery. For a good and efficient battery system, there is always a good ion transport mechanism between the anode and the cathode through the electrolyte and the acquisition and flow of electrons through the external circuit during an electrochemical reaction. Different deposition methods have been used in the design and fabrication of cathode materials for different solid-state battery applications. Production of $LiCoO_2$ cathode materials using radio frequency (RF) magnetron sputtering deposition technique for micro battery applications was studied by Jullien et al. [45], MoO_3 thin-film cathodes using DC sputter deposition technique for Li-ion rechargeable battery applications [39] and V_2O_5 thin-film cathodes using radio frequency (RF) reactive sputtering technique for high-performance thin-film batteries (TFB) [46].

3.4 Deposition technique for anode materials in thin-film batteries

A good anode material should be a good conductor and reducing agent and possess a high electrical energy density. They are made from carbon (graphite), lithium metals, lithium alloys, titanium and its alloys, silicon and other metals. Several methods have been developed on good anode materials for solid-state thin-film batteries for micro-mechanical system applications. Reto Pfenninger et al. [47] studied the electrodeposition of Li-garments on solid state anode. The group worked on metal oxide anodes of $Li_4Ti_5O_{12}$ with good ionic conductivity and less Li-dendrite formation for microbattery applications [47]. Pulsed electrodeposition of Sn-Cu anodes has been studied and developed for enhanced cycle performance in micro battery applications [48, 49]. Spray deposition of silicon anode materials with high energy density for solid-state battery application was developed by Shin Kimura et al. [50]. Many researchers have fabricated different anode materials and their reports have shown that anodes fabricated with good properties have shown to exhibit better battery performances. $Li_4Ti_5O_{12}$ anode thin film was deposited on a magnesium oxide (MgO) substrate using the pulse laser deposition (PLD) technique [47], pulsed laser deposition method was used to fabricate a silicon thin-film anode with iron sulphide for solid-state battery applications [51]. DC magnetron sputtering technique was reported by L. Baggetto et al. for Cu_2Sb thin-film deposition for sodium ion (Na-ion) batteries because of excellent storage capacity, high-rate capacity, good reaction potential and decent cycling capacity retention [52]. Silicon/carbon (SiC) thin-film anode was fabricated using a radio frequency (RF) magnetic sputtering technique with improved cycling performance and high current density for thin-film battery applications.

3.5 Deposition technique for electrolyte materials in thin-film batteries

A solid-state electrolyte is a solid ionic and electron-insulating material that promotes the movement of ions from the cathode to the anode during charging and the moving of ions from the anode to the cathode during discharging conditions. There are different types of electrolytes (glass, crystalline and ceramic) depending on the primary material makeup [53]. Several works have been done on solid-state electrolytes and more researchers keep working to get an electrolyte that will improve the thin-film battery characteristics such as reduced package size, increased safety and enhanced power and energy density [54]. Solid polymer electrolytes made of polyvinylpyrrolidone (PVP) with dopant p-amino benzoic acid (PABA) and poly p-amino benzoic acid (PPA) were developed by Sangeetha et al. for improved dielectric property and ionic conductivity applications [55]. Solid-state electrolytes have been fabricated using pulsed laser deposition (PLD) technique [56, 57], sol-gel techniques [58], atomic layer deposition [59, 60], chemical solution deposition (CSD) [61], etc. for microbattery and other solid-state battery applications.

3.6 Deposition technique for current collector materials

Current collectors are made of copper and/or aluminum foils that collect electrons from the electrochemical reaction of the electrodes and external circuit and at the same time support the material layers of the cathode and the anode. For a material to function efficiently as a current collector, it must be cheap and available, very light material, a good electrical conductor and chemically stable and resistant to corrosion [62]. The importance of current collectors in the solid-state battery (SSB) industry has necessitated research focus on the materials for its fabrication and applications. One-step galvanostatic electrodeposition of lithium in 3D porous copper-based nanoflake structures (3D Cu-NF@Cu foam) as anode current collector for high energy density lithium metal batteries was reported by Yuanyuan Xia et al. [63], electrode polymer-carbon composite current collector foil for bipolar lithium-ion battery applications was reported by Fritsch et al. [64]. Also, Zhen Hou et al. designed a current collector using a nanostructured silver lipophilic layer on a copper foil through an electroless plating electrodeposition process for a stable lithium metal anode [65], fabrication of nickel-phosphorous-modified copper current collector using facile electroless plating electrodeposition method aimed at superior coulombic efficiency for microbattery applications [66], etc.

4. Mechanism

Solid-state thin-film batteries have solid components for the electrodes (cathode and anode) and the electrolyte. They are made by stacking a thin-film electrolyte on the cathode and anode in a vacuum state as shown in **Figure 1**. The principal operation of thin-film batteries works in the same way rechargeable batteries work. The lithium ions migrate from the cathode to the anode and generate electrical energy when the battery is charged and stored in a current collector. During discharge, the lithium ions move from the anode back to the cathode, the electrons move from the cathode to the anode and a current between the cathode and the anode is created as a result of the potential difference between the electrodes [6].

5. Fundamentals of thin-film batteries

Electrodes (both positive and negative) and electrolytes make up the bulk of a current battery's physical components. From an electrical engineering perspective, their properties can be deduced from the first principles. A chemical reaction is what gives a battery its ability to provide electrical energy (the battery converts chemical energy to electrical energy) [67]. Electrodes are required to conduct electrons for electrical energy to be realized in the external circuit. On the other hand, the electrolyte need not. Because if it did, the battery's internals would be accessible to the electrons, rather than just the external circuit, leading to self-discharge and, ultimately, no usable energy at the battery terminals [68]. There is no such thing as an "exceptional" all-solid-state thin-film battery (ASSTFB) [69]. A thin film version of the same three components is used in an ASSTFB, as shown in **Figure 1**.

According to A.G. Buyers [70], the potential energy measured between electrodes for any electrochemical cell, as depicted in **Figure 3**, is the difference in the standard free energy as defined in Eq. (1).

$$\Delta G = -nFE \tag{1}$$

where ΔG = change in free energy, n = number of electrochemical equivalents transferred in the cell reaction, E = measured voltage (EMF) and F = Faraday constant.

For a material to conduct electricity, the charge must be transferred from one location to another. This charge transfer can take the form of electrons, holes (the lack of an electron), ions or ion vacancies [71]. Interstitial hopping mechanisms, such as the Frenkel defect reaction [67], allow metal ions such as Li + to migrate as defined in Eq. (2).

$$M_m^x + V_i^x \rightarrow M_i^\bullet + V'_m \tag{2}$$

Metal ions like Li continue to occupy the same site (denoted by M_m^x), while vacated sites (denoted by V_i^x), newly occupied sites (denoted by $M_i^\bullet$) and vacated sites (denoted by V'_m) are created. Neutral, positive and negative charges are represented by the superscripts x, • and ′. It follows that the electrolytes need to have enough charge carriers for proper ion movement, including that of lithium ions. As a result, according to the definition in (3), ionic conductivity, σ, is proportional to the density of the charge carrier, c.

$$\sigma = q\, x\, \mu\, x\, c \tag{3}$$

where q = charge of the ions and μ = mobility of ions according to the Nernst-Einstein equation, which is defined in Eq. (4).

$$\mu = \frac{D\, x\, q}{kT} \tag{4}$$

From (4) and according to constable [71], it is determined that diffusivities, D, fluctuate with temperature as a thermally activated Boltzmann process defined in Eq. (5)

$$D = D_o e^{-E_d/kT} \tag{5}$$

where k, T and E_d represent the Boltzmann constant, absolute temperature, and diffusivity activation energy, respectively. The interstices should have enough space for the Li + ions to hop around. Metal ion conductivity, in this case, Li+, σ, can be calculated using Fick's second law and the Nernst-Einstein equation.

$$\sigma T = \sigma_o e^{-E_d/kT} \tag{6}$$

σ_o is the pre-exponential constant.

Charge transfer occurs at the interface between the electrodes (cathode and anode) and the electrolyte simultaneously with ion migration through the electrolyte. Therefore, the electrode needs to be highly conductive not only in terms of ions but also electrons. Unfortunately, the electronic conductivity of bare electrode materials is quite low. To provide reasonably good electronic conductivity, some conductive materials such as carbon and carbon derivatives should be incorporated with the electrode materials, and also electrode particles sometimes are coated with carbon layers.

6. Advantages over other battery types

Thin-film lithium-ion batteries offer improved performance due to their higher average output voltage, lighter weights, higher energy density, long cycling life (1200 cycles without degradation) and ability to operate in a wider temperature range (between −20 and 60°C) when compared with the standard lithium-ion batteries [72, 73].

Lithium-ion transfer cells stand out as the most promising systems for meeting the need for high specific energy and high power at a low manufacturing cost [74].

Each electrode in a thin-film lithium-ion battery can accept lithium ions in either direction, creating a Li-ion transfer cell. The components of a battery, including the anode, solid electrolyte, cathode and current leads, must be fabricated into multi-layered thin films using the appropriate technologies to build a thin-film battery [75, 76].

The electrolyte in a thin-film-based system is often a solid electrolyte that can take on the form of a battery. This differs from traditional lithium-ion batteries, which typically use a liquid electrolyte [77]. If the liquid electrolyte is not suitable for use with the separator, its use can be complicated. As a general rule, liquid electrolytes necessitate a larger battery, which is not ideal when trying to get a high energy density in the final product.

Polymer electrolytes, which are commonly used in thin-film flexible Li-ion batteries, can serve multiple functions, including those of electrolyte, separator and binder. Since the problem of electrolyte leakage is thus avoided, flexible systems can be built [78].

Finally, unlike traditional liquid lithium-ion batteries, solid systems can be packed together densely to maximize energy density. Thin-film batteries production have the advantage of high energy densities [79].

7. Application

More and more uses, with varying demands for energy storage, are finding it difficult to stockpile electricity. Printed batteries are only ever mentioned in the literature when the energy demand of a thin-film battery application is less than 1 Ah [5]. Since the battery thickness needed to power a gadget may be drastically reduced with thin-film lithium-ion batteries, slimmer portable electronics can be created with them. These power sources could be used in a wide variety of implantable medical devices, including; Implanted Cardioverter Defibrillators (ICD), Cochlear Implants, Implanted Drug Pumps, Implanted Pacemakers etc.

Thin-film batteries are an efficient means of storing the intermittently produced electricity from solar and other renewable energy sources. It is possible to design these batteries with a negligible self-discharge rate, allowing them to be stored for extended periods without suffering a serious loss of energy capacity [80]. When completely charged, these batteries may also provide more consistent energy for a smart electrical grid [81].

Similar in size to a conventional credit card, “smart cards” actually contain a microchip that can be read, written to, authorized or used to process information. These cards are manufactured in fairly severe environments, with temperatures reaching as high as 150°C. At this severe temperature, other batteries could suffer damage because their internal parts could degas or simply malfunction. Thin- film lithium-ion batteries, however, may operate in temperatures ranging from −40 to 150°C [82]. In addition, the durability of thin film lithium-ion batteries may be advantageous in other applications that involve temperatures that the human body cannot withstand [81].

Radiofrequency identification (RFID) tags are employed in logistics and stock management and are frequently included in discussions of the Internet of Things (IoT) [83, 84]. RFID tags have several applications, some of which include authentication, identification and security. Some RFID tags have built-in sensing technologies that can pick up data about their immediate surroundings. With a higher output battery, the RFID tag may be read from greater distances [85]. As these tags become more sophisticated, the battery demands will have to keep up, and thin-film batteries can be incorporated into these tags due to the battery’s versatility in terms of size and shape, as well as the ability to power the tag’s functions. The disposable applications of RFID technology may be made possible by the low production costs of thin-film batteries [81].

As these tags become more sophisticated, the battery demands will have to keep up, and thin-film batteries can be incorporated into these tags due to the battery’s versatility in terms of size and shape, as well as its ability to power the tag’s functions.

Batteries for implantable medical devices must be reliable and able to give power for an extended period. These batteries must have a low self-discharge rate while not in use and a high-power rate when it is being used, particularly for usage in an implantable defibrillator [86]. Batteries for implantable medical devices should have a high capacity for charge and discharge cycles so that the devices can go longer between servicing and replacement [87]. The power needs of implantable medical devices are ideally met by thin-film batteries. Since the electrolyte in thin-film batteries is solid rather than liquid, they may be shaped in a wide variety of configurations without the risk of leakage, and it has been found that certain types of thin-film batteries can withstand charging and discharging for up to 50,000 times.

The global applications of thin-film batteries cannot be fully enumerated here. They also find applications in smart cards, touch screens, wireless sensors, laptops, astronomical mirrors, photovoltaic energy generation and storage, RFID tags, implantable devices and many others.

8. Conclusion

In this chapter, the overview of types of thin-film batteries, deposition techniques, mechanisms, fundamentals, advantages and applications of thin-film batteries were given. We summarized the types of thin-film manufacturing technologies, and the choice of any technological pathway will depend on the end-use applications of the thin-film batteries. Physical and chemical methods are used to prepare thin-film batteries. These techniques are used to modify the surface properties of the batteries. In chemical vapor deposition, films are deposited from various chemical reactions to generate the vapor that is deposited on the substrate, while the physical deposition technique involves physical mixing and deposition of the mixture on the substrate without any chemical reaction. The need to address the current energy crisis using the most fascinating next-generation energy storage systems and vast applications of thin-film batteries has driven research focus on thin-film batteries. Thin-film batteries have a wide area of applications covering the Internet of Things (IoT), implantable medical devices, integrated circuit cards, smart watches, radio-frequency identifier (RFID) tags, remote sensors, smart building control, astronomical mirrors and other wireless devices.

Acknowledgements

This work was not supported by any fund and was self-funded by the authors.

Conflict of interest

The authors declare no conflict of interest.

Author details

Macdenis Egbuhuzor[1*], Solomon Nwafor[2], Chima Umunnakwe[3] and Sochima Egoigwe[2]

1 Department of Materials and Metallurgical Engineering, University of Nigeria, Enugu, Nigeria

2 Department of Mechatronic Engineering, University of Nigeria, Enugu, Nigeria

3 Department of Agric and Bioresources Engineering, University of Nigeria, Enugu, Nigeria

*Address all correspondence to: onyekachi.egbuhuzor@unn.edu.ng

References

[1] Dai W, Qiao Y, Ma Z, Wang T, Fu Z. All-solid-state thin-film batteries based on lithium phosphorus oxynitrides. Materials Futures. 2022;**1**(3):032101

[2] Liu J, Wang J. New hybrid organic-inorganic thin films by molecular layer deposition for rechargeable batteries. Frontiers in Energy Research. 2021;**9**(296-598X):1-8

[3] Liang X, Tan F, Wei F, Du J. Research progress of all solid-state thin film lithium battery research progress of all solid-state thin film lithium battery. IOP Conference Series Earth and Environmental Science. 2019; **218**(012138):1-15

[4] Ribeiro JF, Silva MF, Carmo JP, Goncalves LM, Silva MM, Correia JH. Solid-State Thin-Film Lithium Batteries for Integration in Microsystems. In: Scanning Probe Microscopy in Nanoscience & Nanotechnology. 2013; **3**(1):575

[5] Clement B et al. Recent advances in printed thin-film batteries. Engineering. 2022;**13**(6):238-261

[6] Critchley L. How thin film batteries work, AZoM. 2018. [Online]. Available from: https://www.azom.com/article.aspx?ArticleID=15815. [Accessed: 18-Sep-2022]

[7] Heckner KH, Kraft A. Similarities between electrochromic windows and thin film batteries. Solid State Ionics. 2002;**152–153**:899-905

[8] Huebner G, Krebs M. Printed, Flexible Thin-film-batteries and Other Power Storage Devices. China: Elsevier Ltd; 2015

[9] Oliveira J, Costa CM, Lanceros-Méndez S. Printed Batteries. 2018th ed. Braga, Portugal: John Wiley & Sons Ltd.; 2018

[10] Kipphan H. Handbook of Print Media: Technologies and Production Methods. 1st ed. Berlin, Heidelberg: Springer; 2001

[11] Gaikwad AM, Steingart DA, Tse Nga Ng, Schwartz DE, Whiting GL. A flexible high potential printed battery for powering printed electronics. Applied Physics Letters. 2013;**102**(23):1-5

[12] Krebs M, Wendler M, Huebner G. Development of printed thin and flexible batteriues. Science and Technology. 2011;**2011**:32-41

[13] Yang P, Li J, Lee SW, Fan HJ. Printed zinc paper batteries. Advancement of Science. 2022;**9**(2):1-9

[14] Gaikwad AM, Ostfeld AE, Arias AC. In: Lanceros-Méndez S, Costa CM, editors. Applications of Printed Batteries, in Printed Batteries. First ed. Califonia: John Wiley & Sons Ltd; 2018. pp. 144-184

[15] Sukkurji PA. Advanced Anode and Cathode Materials for Li-ion Batteries: Application to Printing Methodology. Kochi, India: Karlsruher Instituts für Technologie (KIT); 2021

[16] Sliz R, Molaiyan P, Fabritius T, Lassi U. Printed electronics to accelerate solid-state battery development. Nano Express. 2022;**3**(2):1-6

[17] Cheng EJ, Kimura T, Shoji M, Ueda H, Munakata H, et al. Ceramic-based flexible sheet electrolyte for Li batteries. ACS Applied Materials & Interfaces. 2020;**12**(9):10382-10388

[18] Shomura R, Tamate R, Matsuda S. Lithium-ion-conducting ceramics-coated

separator for stable operation of lithium metal-based rechargeable batteries. Materials (Basel). 2022;**15**(1):1-11

[19] Nestler T et al. Separators - Technology review: Ceramic based separators for secondary batteries. AIP Conference Proceedings. 2014;**1597**: 155-184 February 2015

[20] Heydecke J. Introduction to Lithium Polymer Battery Technology, White Pap. Jauch. Germany: Jauch Quartz GmbH and Jauch Battery Solutions GmbH; 2018. pp. 1-15

[21] R. Batteries. Lithium Polymer Information : Lithium Polymer (LiPo) Batteries , Radio Control Accessories & ... Page 1 of 4 Lithium Polymer Information : Lithium Polymer (LiPo) Batteries, Radio Control Accessories & ... Page 2 of 4, Lithium Polymer Information. 2007. [Online]. Available from: http://www.batteryuniversity.com. [Accessed: 16-Oct-2022]

[22] Long L, Wang S, Xiao M, Meng Y. Polymer electrolytes for lithium polymer batteries. Journal of Materials Chemistry A. 2016;**4**(26):10038-10039

[23] EASE. Nickel Metal Hydride Batteries. Brussels: European Association for Storage of Energy; 2018

[24] Abdin Z, Khalilpour KR. Single and polystorage technologies for renewable-based hybrid energy systems. In: Polygeneration with Polystorage: For Chemical and Energy Hubs. London, UK: Elsevier Inc.; 2018. pp. 77-131

[25] Shin SM, Shin DJ, Jung GJ, Kim YH, Wang JP. Recovery of electrodic powder from spent nickel-metal hydride batteries (NiMH). Archives of Metallurgy and Materials. 2015;**60**(2): 1139-1143

[26] Eckertová L. Methods of Preparation of Thin Films, in Physics of Thin Films. 1977th ed. 1977. pp. 1–38

[27] Geremew T. Thin film deposition and characterization techniques. Journal of 3D Printing and Applications. 2022; **1**(2):1-24

[28] Jilani A, Abdel-wahab MS, Hammad AH. Advance deposition techniques for thin film. In: Nikitenkov NN, editor. Modern Technologies for Creating the Thin-film Systems and Coatings. 2017th ed. London: Intechopen; 2017. pp. 137-149

[29] Velmurugan R, Subramanian B. Physicochemical approaches for thin film energy storage devices through PVD techniques Ramasamy. In: Management and Applications of Energy Storage Devices electrolytes. 1989th ed. Vol. 1989. London: Intech; 1989. pp. 1-28

[30] Shishkin A. Methods of Physical Vapor Deposition. Russia: Mosco Institute of Physics and Technology; 2019

[31] Cho AY, Arthur JR. Molecular beam epitaxy. Progress in Solid State Chemistry. 1975;**10**(PART 3):157-191

[32] Masuda T, Faucher J, Lee ML. Molecular beam epitaxy growth of germanium junctions for multi-junction solar cell applications. Journal of Physics D: Applied Physics. 2016;**49**(46):465105

[33] Kaun SW, Wong MH, Mishra UK, Speck JS. Molecular beam epitaxy for high-performance Ga-face GaN electron devices. Semiconductor Science and Technology. 2013;**28**(7):1-15

[34] Vorobiev A. Pulsed laser deposition of thin films of functional materials. Chalmers Micro- and Nanoprocessing Technologies. 2016;**20**:1-49

[35] Perrière J, Millon E, Craciun V. ZnO and ZnO-related compounds. Pulsed Laser Deposition of Thin Films: Applications-Led Growth of Functional Materials. 2006:261-289

[36] Schey B. Pulsed laser deposition of high-temperature superconducting thin films and their applications. In: Eason R, editor. Pulsed Laser Deposition of Thin Films: Applications-Led Growth of Functional Materials. First ed. Augsburg, Germany: Wiley-VCH Verlag GmbH & Co. KGaA; 2006. pp. 313-331

[37] Mooney JB, Radding SB. Spray pyrolysis processing. Annual Review of Materials Science. 1982;**12**:81-101

[38] Ubaidah SN, Ying KK, Khuan NI. Electrodepostion: Principles, applications and methods. Industrial Technology Division. 2011:1-5

[39] Methani W et al. Nanomechanical, structural and electrochemical investigation of amorphous and crystalline MoO3 thin-film cathodes in rechargeable li-ion batteries. Batteries. 2022;**8**(80):1-16

[40] Kim K, Raymond D, Candeago R, Su X. Selective cobalt and nickel electrodeposition for lithium-ion battery recycling through integrated electrolyte and interface control. Nature Communications. 2021;**12**(1):1-10

[41] O'Sullivan EJ. Electroless Deposition, Electrochemical Society. USA: Electrochemical Society; 2011

[42] Oviroh PO, Akbarzadeh R, Pan D, Coetzee RAM, Jen TC. New development of atomic layer deposition: Processes, methods and applications. Science and Technology of Advanced Materials. 2019;**20**(1):465-496

[43] Bokov D et al. Nanomaterial by Sol-Gel method: Synthesis and application. Advances in Materials Science and Engineering. 2021;**2021**:1-21

[44] Bahuguna RSG, Kumar A. Thin film coating through sol-gel technique. Chemical Sciences. 2016;**6**. no. September 2017

[45] Julien CM, Mauger A, Hussain OM. Sputtered $LiCoO_2$ cathode materials for all-solid-state thin-film lithium microbatteries. Materials. 2019;**12**(2687): 1-27

[46] Jeon EJ, Shin YW, Nam C, Il Cho W. Characterization of all-solid-state thin-film batteries with V_2O_5 thin-film cathodes using ex situ and in situ processes. Journal of the Electrochemical Society. 2001;**148**(4):318-322

[47] Pfenninger R, Afyon S, Garbayo I, Struzik M, Rupp JLM. Lithium titanate anode thin films for Li-Ion solid state battery based on garnets. Advanced Functional Materials. 2018;**28**(21): 1800879

[48] Beattie S, Dahn J. Single bath, pulsed electrodeposition of Copper-Tin alloy negative electrodes for lithium-ion batteries. Journal of the Electrochemical Society. 2003;**150**:894-A898

[49] Hu R, Liu H, Zeng M, Liu J, Zhu M. Progress on Sn-based thin-film anode materials for lithium-ion batteries. Chinese Science Bulletin. 2012;**57**(32): 4119-4130

[50] Kimura S, Sakabe J, Mitsuishi K, Ohnishi T, Takada K. Anode properties of Si nanoparticles in all-solid-state Li batteries. ACS Applied Energy Materials. 2019;**2**(10):7005-7008

[51] Miyazaki R. High-capacity anode materials for all-solid-state lithium batteries. Frontiers in Energy Research. 2020;**8**(171):1-6

[52] Baggetto L, Allcorn E, Manthiram A, Veith GM. Electrochemistry Communications Cu 2 Sb thin fi lms as anode for Na-ion batteries. Electrochemistry Communications. 2013;**27**:168-171

[53] Xia H, Wang HL, Xiao W, Lai MO, Lu L. Thin film Li electrolytes for all-solid-state micro-batteries. International Journal of Surface Science and Engineering. 2009;**3**(1–2): 23-43

[54] Ihlefeld JF, Clem PG, Doyle BL, Kotula PG, Fenton KR, Apblett CA. Fast lithium-ion conducting thin-film electrolytes integrated directly on flexible substrates for high-power solid-state batteries. Advanced Materials. 2011;**23**(47):5663-5667

[55] Sangeetha S, Kalaignan GP. Tribological and electrochemical corrosion behavior of Ni–W/BN (hexagonal) nano-composite coatings. Ceramics International. 2015;**41**(9): 10415-10424

[56] Siller V, Morata A, Arenal R, Gonzalez-rosillo JC, Juan Migue L'opez del Amo. High performance LATP thin film electrolytes for all-solid-state microbattery applications. Materials Chemistry A. 2021:1-10

[57] Kuwata N, Iwagami N, Tanji Y, Matsuda Y, Kawamura J. Characterization of thin-film lithium batteries with stable thin-film Li_3PO_4 solid electrolytes fabricated by ArF excimer laser deposition. Journal of The Electrochemical Society. 2010;**157**(4): A521-A527

[58] Zheng ZFF, Song SDD, Wang Y. Sol-gel-processed amorphous lithium ion electrolyte thin films: Structural evolution, theoretical considerations, and ion transport processes. Solid State Ionics. 2016;**287**(2016):60-70

[59] Liu J et al. Atomic layer deposition of lithium tantalate solid-state electrolytes. Journal of Physical Chemistry C. 2013; **117**:20260-20267

[60] Aaltonen T, Alnes M, Nilsen O, Costelle, Fjellvag HL. Lanthanum titanate and lithium lanthanum titanate thin films grown by atomic layer deposition. Journal of Materials Chemistry. 2010;**20**:2877-2881

[61] Oh E et al. Extremely thin bilayer electrolyte for solid oxide fuel cells (SOFCs) fabricated by chemical solution deposition (CSD). Advanced Materials. 2012;**24**:3373-3377

[62] Luo S, Pei K. Construction and modification of copper current collectors for improved Li metal batteries. In: Lithium-Ion Batteries - Recent Advanced and Emerging Topics. 2022nd ed. London: Intechopen; 2022. pp. 1-15

[63] Xia Y et al. Application of electrodeposited Cu-metal nanoflake structures as 3D current collector in lithium-metal batteries. Nanotechnology. 2022;**33**(24):245406

[64] Fritsch M et al. Lightweight polymer-carbon composite current collector for Lithium-Ion batteries. Batteries. 2020;**12**(60):1-18

[65] Hou Z et al. Lithiophilic Ag nanoparticle layer on Cu current collector toward stable Li metal anode. ACS Applied Materials & Interfaces. 2019;**11**(8):8148-8154

[66] Cai K et al. Facile electroless plating method to fabricate a Nickel-Phosphorus-modified copper current collector for a lean

lithium-metal anode. ACS Applied Materials & Interfaces. 2022;**14**(40): 45433-45443

[67] Wu T, Dai W, Ke M, Huang Q, Lu L. All-solid-state thin film μ-batteries for Microelectronics. Advancement of Science. 2021;**8**(19):2100774

[68] Hooper A, McGeehin P. Physical principles of advanced battery design. Physics in Technology. 1981;**12**(2):45-53

[69] Dudney NJ, West WC, Nanda J. All solid-state thin film batteries. Handbook of Solid State Batteries. 2015;**10**:593-625. no. 1142/9487

[70] Buyers AG. Galvanic cell composed of refractory metal electrodes and a calcia-stabilized zirconia electrolyte. Journal of the American Ceramic Society. 1965;**48**(3):122-125

[71] Constable S. Geomagnetic induction studies. In: Treatise on Geophysics. Vol. 2015. USA: Elsevier; 2015. pp. 219-254

[72] Tomaszewska A et al. Lithium-ion battery fast charging: A review. eTransportation. 2019;**1**:100011

[73] Bashir T et al. A review of the energy storage aspects of chemical elements for lithium-ion based batteries. Energy Materials. 2021;**1**(2):100019

[74] Eshetu GG et al. Production of high-energy Li-ion batteries comprising silicon-containing anodes and insertion-type cathodes. Nature Communications. 2021;**12**:1-14

[75] Nzereogu P, Omah A, Ezema FI, Iwuoha E, Nwanya A. Anode materials for lithium-ion batteries: A review. Applied Surface Science Advances. 2022; **9**:2022

[76] Xu L et al. Interfaces in solid-state lithium batteries. Joule. 2018;**2**(10): 1991-2015

[77] Ribeiro JF, Silva MF, Carmo JP, Gonçalves LM, Silva MM, Correia JH. Solid-state thin-film lithium batteries for integration in microsystems, in scanning probe microscopy. In: Bhushan B, editor. Nanoscience and Nanotechnology 3. Berlin, Heidelberg: Springer-Verlag; 2013. pp. 575-619

[78] Wikipedia. Thin-film lithium-ion battery. en.wikipedia.org/wiki?cv=1&curid=19236411. 2022. [Online]. Available from: https://en.wikipedia.org/wiki?cv=1&curid=19236411. [Accessed: 22-Sep-2022]

[79] Cui G. Reasonable design of high-energy-density solid-state lithium-metal batteries. Matter. 2020;**2**(4):805-815

[80] Seong WM et al. Abnormal self-discharge in lithium-ion batteries. Energy & Environmental Science. 2018; **11**(4):970-978

[81] Brett S. Applications of thin film batteries. AZoM. 2018. [Online]. Available from: https://www.azom.com. [Accessed: 07-Sep-2022]

[82] Nagasubramanian G, Doughty DH. Electrical characterization of all-solid-state thin film batteries. Journal of Power Sources. 2004;**136**(2 SPEC. ISS):395-400

[83] Laxmi AR, Mishra A. RFID based logistic management system using Internet of Things (IoT), in 2018 Second International Conference on Electronics, Communication and Aerospace Technology (ICECA). 2018. pp. 556–559

[84] Tan WC, Sidhu MS. Review of RFID and IoT integration in supply chain management. Operations Research Perspectives. 2022;**9**:100229

[85] Profetto L, Gherardelli M, Iadanza E. Radio frequency identification (RFID) in health care: Where are we? A scoping review. Health and Technology. 2022; **12**(5):879-891

[86] Bock DC, Marschilok AC, Takeuchi KJ, Takeuchi ES. Batteries used to power implantable biomedical devices. Electrochimica Acta. 2012;**84**: 155-164

[87] Khanna VK. Batteries for Implants, in Implantable Medical Electronics. 2016th ed. Cham: Springer International Publishing; 2016. pp. 167-183